SHOULDER OF MUTTON FIELD

The Retail Butcher's Trade in Camden

To John
Best Wishes
Des.

Des Whyman

Nottingham University Press
Manor Farm, Main Street, Thrumpton
Nottingham, NG11 0AX, United Kingdom

NOTTINGHAM

First published 2010

British Library Cataloguing in Publication Data
Shoulder of Mutton Field - The Retail Butchers Trade in Camden
Whyman, D.

ISBN 978-1-907284-73-1

Disclaimer

Every reasonable effort has been made to ensure that the material in this book is true, correct, complete and appropriate at the time of writing. Nevertheless, the publishers and authors do not accept responsibility for any omission or error, or for any injury, damage, loss or financial consequences arising from the use of the book.

Typeset by Nottingham University Press, Nottingham
Printed and bound by Lightning Source, England

CONTENTS

ACKNOWLEDGEMENTS

One of the great pleasures during the course of research and compiling the material for this book has been the encouragement from my wife Vicki, family, friends, trade colleagues and new acquaintances. In particular the generous help from archivist and friend Mark Aston at the Camden Local Studies and Archive Centre.

The names in the following lists in no particular order are those to whom I also owe a debt of gratitude. For those anecdotal story tellers for whom life has completed its cycle I have faithfully kept to the spirit of their recorded memories. In the dreadful event that I have inadvertently not credited a name I offer my sincere apologies in the hope my oversight will be forgiven.

Camden Local Studies & Archives Centre.
Mark Aston
Aidan Flood
Malcolm Holmes, M.B.E.
Richard Knight
Lesley Marshall

Camden History Society and Essay contributors 1980, 1983.
Mrs Ashbolt
Ena Baker
G.L. Evans
Reginald Pleeth
Mrs Reeks

Butchers Shop Proprietors.
Ahad Dasht - The Pure Meat Co.
Bob Enright – Barretts Traditional Butchers
Tom and Paul Buttling - T. Buttling. (Devon)
Maou Ghosseiri - Jackson Bros.
Lee and Phyllis Harper - Highgate Butchers.
Martin Leahy -Elite Meats.
Philip F. Cramer - Cramers Butchers

The Worshipful Company of Butchers.
Collin Cullimore, C.B.E. DL
Michael Katz, M.B.E.
Fred Mallion, M.B.E.
Henry Tattersall
John Tuckwell

Consultant Archivists, Unilever plc.
Mr A. A. Cole, Sophie Clapp

The Sainsbury Archives.
Jayne Stewart, Bridget Williams

Dewhurst (Butchers) Ltd. Estate Office
Mrs E. de Borg

Rotary Club of St. Pancras.
Chris Leverton

University of Glasgow.
Elaine Mackay

INDIVIDUALS

Austin Burnett
Bob Boud, (Australia)
Lord Donald Coggan and Lady Jean Coggan
Patrick and Phillip Coggan
Dennis Cole
James Dunning
Ted Ford
Ron Hardy, (Canada)
Paul Langley
Geoff Linsay
Mrs Emily Matthews
Simon Morris
John Norris
Anton Obrist, (Switzerland)
Kenneth Page
Wendy Trewin
E.G.Turner
Gladys Saunders
Ernest Vincett
Bennet Wright

ILLUSTRATIONS

The majority of illustrations including the cover pictures for this book with the exception of those from the following individuals, organisations and institutions listed below are from the Camden Local Studies and Archives collection. I wish to express my thanks to them all for the opportunity to reproduce them in this publication.

Mr E. Ford: 30

Mr T. Buttling: 55

G. Brazil & Co Ltd: 85

Mr P. Cramer: 78, 118

Ambrose Keevil: 123

Unilever plc Archives: 42

National Railway Museum: 98

Guildhall Library, City of London: 14

City of Westminster Archive Centre: 50

The Worshipful Company of Butchers: 12, 47

Times Newspaper: 39, 72, 87, 113

Coggan Family: 102, 104, 105, 106

London Metropolitan Archives: 83, 90

Cooperative Society Archives: 97

The Sainsbury Archive, Museum of London, Docklands: 122, 126, 127

National Archives: 37, 66, 75, 103, 114, 119, 120, 121

Meat Trade Journal, William Reed Business Media: 24, 35, 38, 42, 57, 68, 95, 125

Author's Collection: 21, 34, 49, 64, 79, 108, 110, 124, 129-137

Dedicated to James Dunning

Who showed by example there is a story to be told.

INTRODUCTION

Shoulder of Mutton Field in Kentish Town, from whence the title of this book is taken, is representative of countless fields throughout the land; upon which from time immemorial animals have been reared for the production of food. From this seemingly simple act of human survival, there has developed through the ages a multitude of services and industries. From scientists to seamen, across the great oceans to the dusty outback of Australia and the lush green pastures of New Zealand to the vast plains of America and the pampas of Argentina, all have contributed to the history of the retail meat trade.

The history of the urban butcher's shop is comparatively modern, and whilst particular districts in the Borough of Camden created in 1965 and consisting of the former Metropolitan Boroughs of Hampstead, St Pancras and Holborn may be unfamiliar to some readers, their growth followed similar patterns elsewhere. It is not possible to profile every butchers shop and therefore Kentish Town has been chosen as representing the development of the trade. The story is not solely confined to statistics or even great events, but just as importantly the employees and customers, essential to their success. It was a time when meat was only second in importance to bread and customers regularly visited the retail butchers shop three or four times a week. Since that era events have decreed otherwise, and this particular style of shopping for meat, together with employment opportunities afforded by the retail butchers trade, has almost passed into history.

The early arrivals could not have foreseen the social changes and technological advances in food retailing that have significantly altered the public's shopping habits, which consequently lead to the closure of literally thousands of butcher's shops nationwide. Most of the shops featured have ceased trading and their various proprietors long since passed away, their premises used for other trades and purposes. Others have been demolished during rebuilding programs that have taken place. For the overwhelming majority, their business activities passing into history unrecorded.

However with a few exceptions they had a common goal, to perfect their craft and give good service, and of course receive a reasonable financial return for their efforts. Those with greater entrepreneurial spirit and good fortune expanded, most were satisfied in their achievements and content in the knowledge their efforts provided an important service within their community. The contents do not purport to be a definitive record of any particular business, rather a readable account of facts and associated material. Within the reader I hope will be pleasantly surprised to discover there is more to the retail butchers trade than a striped apron and two lamb chops.

BUTCHER

The Butchers generally require more Skill to learn their Trade than any other of the Victualing Branches we have mentioned. They muſt not only know how to kill, cut up, and dreſs their Meat to Advantage, but how to buy a Bullock, Sheep, or Calf, ſtanding: They muſt judge of his Weight and Fatneſs by the Eye; and without long Experience are often liable to be deceived in both. Butchers are neceſſary; yet it is almoſt the laſt Trade I ſhould chuſe to bind a Lad to. It requires great Strength, and a Diſpoſition no ways inclinable to the Coward: A Lad may be found about Fourteen or Fifteen. The Wages of a Journeyman is not much more conſiderable than that of a common Labourer.

R. Campbell

THE LONDON TRADEMAN, 1747

1

OPEN FOR BUSINESS

Three centuries ago, in the year 1702, the presence of a hawking meat seller in Kentish Town was recorded in the parish registers; one of the hoards of peddlers and such like that roamed the countryside selling their wares. A solitary figure of simple motives, other than to earn his daily keep, he was observed 'up and down Kentish Town Road' plying his trade. Yet from this fragment of meat trade history in Kentish Town would evolve a concentration of butchers' shops to rival, and at periods exceed, those of its near neighbours in the Borough of Camden.

To what extent, if any, our itinerant tradesman faced competition from retail butchers shops already established in the area by this period is unclear, except to state the geographical position of Kentish Town Road as part of the network of highways northwards from London would seem to suggest the first butcher's shops may have originated in part to serve the needs of passing travellers. Although just how this came about is again uncertain, for the sequence of events may not be as generally imagined. It was then commonplace to begin in business without capital and under the most primitive of circumstances; a cottager selling pickled pork produced from his pigs or the local farmer disposing of surplus meat from animals unsold at market, the tavern landlord selling unused meat or a hawking meat seller upending a barrel by the roadside. All these events and more have led to the eventual development of fixed premises, the most common method being to convert a private dwelling house by simply using the front room and window with perhaps the addition of a canopy that developed into a full front extension to the house. In general, incoming butchers nearly always opted for existing butchers premises to avoid problems with the local authority and unnecessary expense. To these considerations maybe added the sparse population in pre-urban village settlements and thereby a restricted amount of trade, which encouraged business minded inhabitants to acquire an additional source of income. This is to say, the butcher may also be the inn or tavern keeper, farmer, baker, grocer, cheesemonger, a situation extended into any number of combinations and occupations. We may also include the ever present vulnerability of financial failure and incidence of crime, associated or otherwise, that varied with the prevailing economic and social climate.

Of the earliest retail butchers in Kentish Town c1766 was Thomas Hale trading in Old Chapel Row, adjacent to the Jolly Anglers public house on the corner of present day Anglers Lane. Below the apex of the building is a plaque dated 1731, that if relevant to commencement of tenure would place him there many years prior. The name Hale is in evidence before the seventeenth century with the involvement of Henry Hale, albeit by proxy with other parties in the reversion of lands in Kentish Town; and in 1686 a William Hale, in association with Yeoman Thomas Smith of Kentish

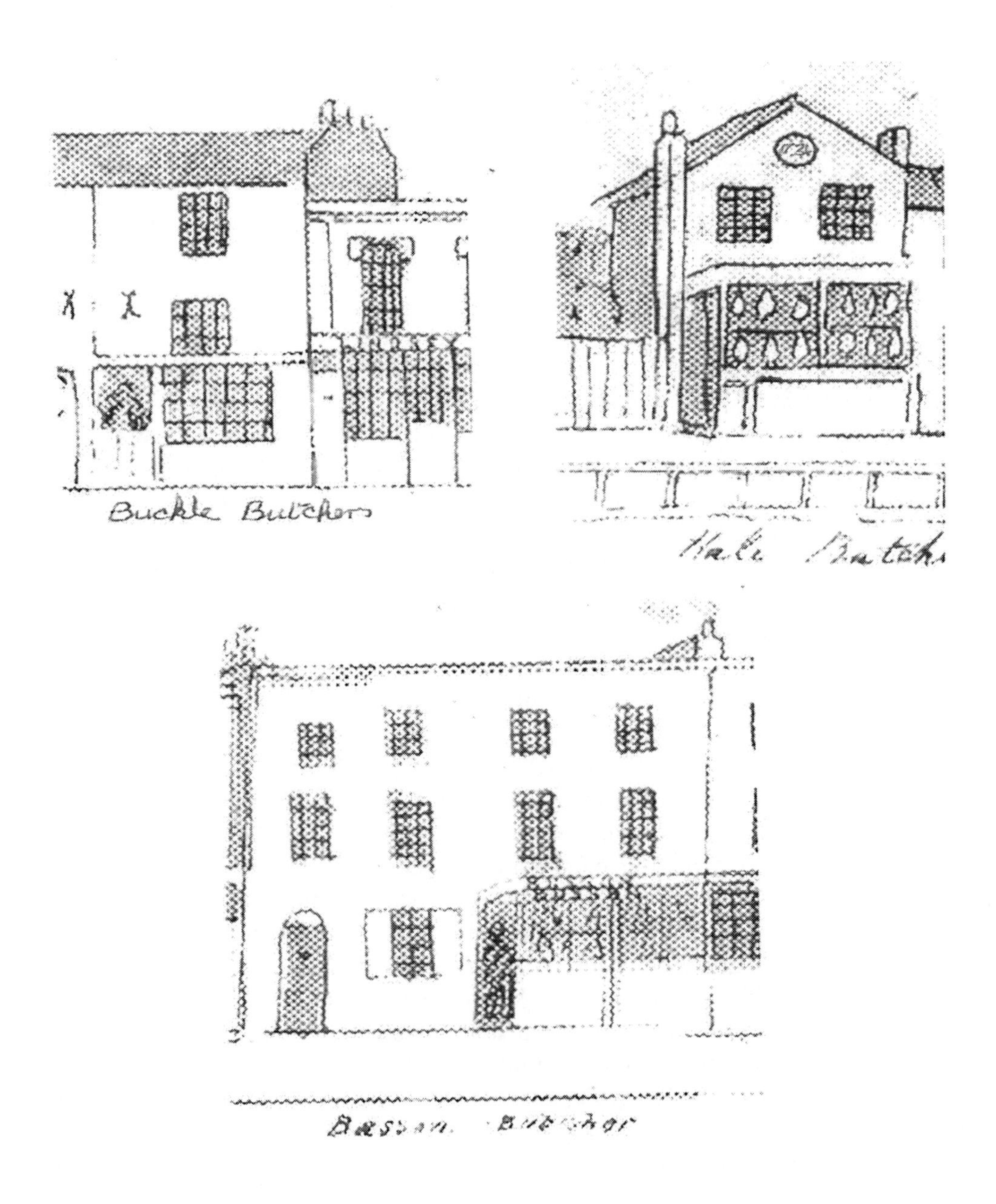

Figure 1: Kentish Town Butchers Shops (1766 - 1846) C/A
L to R, Buckle, Hale, Busson.

Town, with dwelling house, great barn "tilne kilne" and lands Oakfield, Middle Field, Neither Field, and Park Gate Field. This aside there is irrefutable evidence that Thomas Hale the butcher was established in the area by at least 1766, for on 17th December in that year he appeared as witness for the prosecution against assailants William Johnson and William Walker indicted for armed Highway Robbery; at which time during the trial at the Old Bailey he confirmed his trade and address, and also his courage in together with William Jarvis apprehending the culprits during a violent struggle in the Jolly Anglers ale-house. The two robbers, Johnson and Walker were found guilty of violently assaulting and putting Mr Jarvis in fear of his life and were sentenced to death. Many years later in 1781 whilst working in his shop, community spirited Thomas Hale again acted with alertness by chasing and apprehending a suspected housebreaker. The culprit, John Gladman one of three young men involved all appeared for trial at the Old Bailey in which Hale gave evidence. The family feature again six years later when sons, William and Joseph Hale are recorded at the same court. This time they gave evidence for the prosecution against John Walby accused of feloniously stealing; a table cloth, two shirts and one silk handkerchief. In evidence William stated the items were stolen from his garden opposite his shop, his statement carrying substantial weight against the weak explanation from Walby that he found the items. The accused was found guilty and sentenced to be publicly whipped and imprisoned for six months.

By the eighteen hundreds later generations of the family, with confusingly identical forenames, had opened a grocery and cheesemongers shop trading from Fitzroy Place, in addition to the butchery business. The family may have also provided a local coach service because for many years' coachmen and grooms were employed for vehicles far in excess of personal need. Although unconfirmed, there also exists a distinct possibility that other butchers shops recorded as Hale in Hornsey Lane, Marylebone Street, Park Road, Regents Park and Harrington Square, Mornington Crescent in the 1830s were family related. Nevertheless the Kentish Town enterprise, by then a prosperous business and safeguarding their interests with insurance policies registered annually with the Sun Fire Office. In February 1851 the grocery and cheesemongers business, hitherto a partnership between William and Joseph Hale was dissolved by mutual consent with Joseph carrying on the business on his own account. A decision without any adverse effect as Kentish Town born Joseph continued into the eighteen-sixties, at which period was trading and living with his wife Elizabeth and Gwendolyn their eight year old daughter at 6 Upper Fitzroy Place, Later generations of provisions merchants W A. & H Hales trading from 161 Kentish Town Road. The name Hale remained prominent in the food retailing history of Kentish Town, spanning nearly two centuries, from various locations as butchers, tripe dressers and grocers.

By the opening years of the 19th century we see a gradual increase in the number and variety of retail traders. The family concern of Edward and Phillip Wilson, had established adjacent shops trading as traditional and pork butchers between the Old Vine and Bull & Gate public houses in Highgate Road. In company with Joseph Holland sited where Kentish Town library now stands wherein on the 26th of April 1805 he was insured the Sun Insurance Office.

Other retail butchers had made an appearance by 1820 that with a stretch of imagination may still be described as village butchers, these were: Charles Davies, William Hill, Rozea Thomas, and Giles Silverside. Of these we may temporarily exclude the shop belonging to Silverside, meat retailer and wholesaler discussed in detail later and likewise elusive Rozea Thomas, for whom research has thus far proved unproductive.

In James Pigots 1823 London & Provincial Directory buildings and population had grown sufficiently in number for the preamble writer to state: 'Kentish Town is three miles north of London on the road to Highgate and near Hampstead. It is a place of considerable business and has of late years been much improved by the erection of several handsome edifices, and by becoming the residences of many persons of consequences; as here you have the salubrious air of Highgate, without the necessity of climbing the hill. There is a handsome chapel of ease to the parish'.

Figure 2: Old Chapel Row, Kentish Town Road, 19th Century.

The claims of the unknown writer fully justified, as both private and business people were eager to move into the area. A newly erected six roomed house (built as a shop) in Winchester Place, centred in the high street sold within days. As did four houses with shops adjoining the Castle Tavern further south in Kentish Town Road. The presence in 1827 of Thomas Wells, carpenter and dealer in shop fixtures at 11 Mansfield Place, verified a need for such services in Kentish Town and surrounding areas.

There were a considerable number of businesses that for a variety of reasons did not appear in directories such as Christopher Palles, knowledge of whom noted in the annuals of crime in December 1840 as a victim of thievery. The precise location of his butchers shop in Kentish Town is thus far unknown.

Included in Pigots 1823 directory, butcher William Hill who predates this entry by thirteen years and possibly more. He rented a field near his house in Kentish Town in which he grazed some thirty sheep and lambs, from where on the night of 11th August 1813 one sheep valued at 30 shillings was stolen.

The theft of animals, then a capital offence did not bode well for James Saunders and William Thomas who were both caught in a field opposite Church Row the following morning with portions and entrails of the dissected sheep in their possession. At the Old Bailey trial a month later both Saunders aged fifty-six and Thomas aged sixty-four gave implausible stories in their defence resulting in guilty verdicts, and both were sentenced to death.

Figure 3: Charles Davies 287 Kentish Town Road.

Of interest Charles Davies, who lived above his shop at 287 Kentish Town Road, with wife Alice, his butcher son Charles, grandson Harry, niece Jane Clack and three female servants. Formerly known as Old Chapel Row, he shared this section of road in 1867 with nearby trades' people: dairyman Thomas Park, tobacconist, Mrs Womack, John Burge a boot maker and publican Henry Cooper of the Old Farm House on the corner of Holmes Road. Although only part of his shop is visible , it nonetheless adds a physical sense of existence as does the old women sitting on his shop step; that by familiarity may not have been the first time and whom we may hazard a guess is Mrs Womack proprietor of the adjacent tobacconist shop. His presence in Kentish Town first noted in 1823, Charles Davies also traded from number 117 in the same road and a branch shop in the High Street, Somers Town, he also managed a life outside business hours far from uneventful. To begin however, the common practice whereby shopkeepers accepted newspaper advertisement postal replies on behalf of customers. One example of many reveals the career aspirations of women in the eighteen- twenties were far from un-imaginative before emancipation.

> "To families going abroad, a young woman wishes for a situation in any part of the world, but prefers the West Indies, excellent character for honesty and sobriety, no office keeper need apply."
>
> Address post paid. To Mr C. Davies, butcher of Green Street, Kentish Town

Two years prior Charles Davies had displayed the virtue of the Good Samaritan, after a middle aged gentleman fell from his horse outside his shop at the entrance to Kentish Town from Hampstead. He immediately ran to his aid, and with the help of two neighbours carried the man to the side of the road. The horse rider Mr Cooper was fatally injured and at the inquest held later the coroner Mr Stirling's verdict "Died by the visitation of God" was recorded. His good offices were called upon again during Christmas 1825, when John Devkin the local watch house keeper (a rate paying community officer) required his expertise to identify the remains of a sheep stolen by Henry Pitham, who was indicted for wilfully and feloniously killing a sheep, valued at 20 shillings with intent to steal. The eighteen year old defendant was found guilty and sentenced to death.

In December 1843 Charles Davies attended a coroner's inquest for a wholly different reason, causing considerable notoriety for himself and other family members.

The circumstances involved the death of a young man named George Gray, who died from injuries sustained during a prize-fight in Kent. At the inquest held at the town hall in Gravesend and as events transpired proceedings of lengthy duration, the following details emerged.

The prize fighting opponent of the deceased was nineteen year old Henry Ball a bricklayer who lived in Macclesfield Place, Kentish Town. The part played in the proceeding by Charles was to identify the deceased and explain particulars of how his son Alfred, then in police custody for having attended the fight, came to be there and other relevant facts.

The coroner Mr Carttar adjourned the case for more witnesses to be called and bound over Mr Charles Davies in the considerable sum of 200 guineas for his son's appearance on the adjourned day. The coroners sombre closing remarks that taking the most lenient view the case was applicable to murder or manslaughter. The final outcome frustratingly undiscovered, albeit Charles and Alfred were never directly involved. In the remaining years of the eighteen-seventies the Davies shop was acquired by Thomas Boreham, followed in succession by three more butcher proprietors until in c1900 a change of trade as Ernest Almond linen drapers. The other branch in this road in the ownership of his son Charles junior was eventually sold to butchers, the Elvidge brothers in the eighteen-nineties as a going concern.

EXPANSION

From the 1840s the expansion of Kentish Town began in earnest, creating demand and opportunity for a succession of butchers; bringing with them their less appealing trade practices, principally slaughtering and processing of by-products. A way of life that would partially contribute to the less than favourable reputation of Kentish Town; indolent landlords, chronic overcrowding; anti-social behaviour, cowsheds and railways would do the rest. The number of individual males above twenty years of age engaged as butchers in the Metropolis increased by 32% from 4,332 in 1831 to 5,710 in 1841.

In butchers Buckle at Hawley Place and Busson sited in Old Chapel Row, we see examples of dwelling houses used other than for their intended purpose. The general problem of crime never far away, in the winter of 1842 Benjamin Buckle who lived in Hampstead came within hours of loosing his grey delivery pony, which had been stolen by Henry Collins and sold to George Watts a horse slaughter in Maiden Lane, Battle Bridge; the ancient lane then roughly on the line of Brecknock Road and York Road, finishing at the southern end at Battle Bridge, the former name of Kings Cross. His business neighbour in Kentish Town, Richard Bussan, in the unenviable juxtaposition two years later of being onlooker and victim of shoplifter Thomas Sneed who stole 1lb. 6oz weight of mutton valued at 9d. The theft occurred on the evening of 27th November 1846 as Richard returned home from London and saw Sneed take the mutton from the shop-board and conceal it under his jacket. An inveterate petty criminal, when challenged he denied stealing and had the audacity to ask for "Three pennyworth of meat pieces", found guilty at his trial he was sentenced to be confined for two months. The drawings of both shops depicted in a recently discovered small sized panorama, the artist known only by the initials H.G.

We see again the beginning of an extension attached to the unknown butchers premises in York Place depicted on the front cover of this book. This view is taken from the much larger Kentish Town Panorama, an important source of research drawn post 1848, by local artist James King towards the end of his life. He lived from 1837 until his death in 1855 at number three Montague Place with his wife Mary who died in her sixty-ninth year in 1856; the York Place premises at a later date in the 1860s owned by butcher Richard Steven

Figure 4: Robert Balch, 223 Kentish Town Road, 1903.

Morris, then number one, beginning a row of ten houses initially cultivated by retired occupants with independent means. During his stay, renumbered and renamed 317 Kentish Town Road becoming the address of several well known retail meat traders. He vacated the premises to live in retirement in Guildford, Surrey where he died on the 26th February 1871, and his son Richard the sole executor of his fathers will.

We can tentatively assume butchers William Edycoombe and William Chipperfield, amongst the vanguard by 1850 and originally Devon and Hertfordshire men, were in competition from sites in Morton Terrace in Kentish Town Road. The female domestic servants that both employed of valuable assistance to their wives and children. The eldest of the two wives Hanna, married to Chipperfield, having work outside the home as an artist in wax modelling.

It should not be assumed every shop built was an afterthought. For example, in the approach road to Old London Bridge on the east side at the northern end stood 'The Piazza' c1745 specifically constructed to accommodate shops and designed by George Dance the Elder, the City surveyor. Considered by other architects of the day to be a rather uninspired plain block design, save its redeeming feature a colonnade in front of the eight shops, it became known as the piazza. On occasions permission to build shops was refused The vestry dug their heels in concerning a vacant triangular plot of land in Great College Street, (now Royal), after the trustees announced in 1877 their intention to build a number of one-storey shops. The trustees backed down after legal proceedings were commenced, but only after the vestry agreed to use the land as a public highway or an enclosed garden. Acting swiftly to avoid further acrimony, the College Gardens were laid out and dedicated to the public on July 19th of that year.

The initial development of high road and side streets shops was a haphazard affair; there were few if any by-laws against it. All along the main highway, through Kentish Town as elsewhere, houses intended exclusively for private dwelling were being altered to incorporate shops either within or as an extension of the building, Each one according to the occupants' whims and fancies with scant regard to utility or hygiene, resulting in a strange variety of structures and sizes more in keeping with so called shantytowns. Functional rather than ascetic, re-fronting became the most common method. This involved adding a front extension onto the house, sometimes at the gardens expense and relegating the front parlour to that of kitchen or storeroom, with possible use of a yard in the rear; these became known as front, middle and back premises, in practical butchery terms the shop, preparation room and slaughtering area. A run of present-day buildings opposite Kentish Town library were converted in this way. Experience taught butchers to choose the coolest side of the road with a northerly outlook wherever possible to avoid the sun during the hottest part of the day.

Figure 5: Price Bros, Meat Stall, Kentish Town Road, 1903.

Of paramount importance for butchers was choosing a site with easy access for animals, either via a mews behind the shop or an alley from a side street which is why butchers most favoured a corner location. Alternatively, the provision of a cart-way with rooms over the top forming an archway solved the problem. Failing this, with no side entrance, animals came straight through the shop to a slaughtering yard behind; the proverbial bull in a china shop, and more often than not the only way in for proprietors or tradesmen living on the premises. An inherent disadvantage that had unforeseen consequences in future years with hundreds of vacant empty rooms which could not be let to needy families, because prospective business occupiers objected to sitting tenants on the grounds of safety, security and access.

Before the nineteenth century the size of shop was restricted to the boundary of the buildings they fronted, in many instances the width scarcely exceeded twelve feet A typical example and still commonly seen today is one of two butchers' shops owned by Robert Balch in Kentish Town high street. Born into trade, his father Thomas James Balch, owned an open fronted butchers shop in Lamb's Conduit Street, Holborn, throughout the early eighteen hundreds that by official accounts was prone to petty criminals stealing meat on display. His own shop frontage overshadowed by a much larger building advertising the Circulating Library within and as the title implies, a reader's chargeable book lending service provided by Muddie's; the proprietors and shop trade at this time being the Murrell Brothers, stationers. To the left is F. Jones & Son, boot and shoe makers and agents for numerous other enterprises.

The Balch family business at this address date's from c1862, then named Old Chapel Row when a confusing array of sub-division names were routinely used for the shortest stretches of buildings. The family acquired a second shop in the area from butcher Guy Smith at number 144 situated on the eastern side of this road. Robert Balch died on the 15th September 1896 at the family home nearby at number 15 Gaisford Street, he lived there with his wife Sophia and their six children, two of whom Robert and Alfred carried on the business after his death. There is no record of any domestic staff resident in the house, albeit daily help may have been employed and lived elsewhere. The Balch trade was buoyant enough at this period to require the full time services of three men and two boys; however by the outset of the First World War both shops had ceased trading. There existed locally at this time Balch, Balch and Turham at 175a Kentish Town Road, auctioneers and estate agents. It may be there is a family connection between the two businesses, although in 1913 Robert C. Turnham is listed in bankruptcy proceedings by creditors in connection with this firm. However, the family trading name still survives today in the respected firm of Balch & Balch, estate agents.

Further south in eighteen fifty-five, the shops of Thomas Josling and James Holt in Hawley Place, then opposite present day St Andrews Greek Church, had been built in the same manner; although Frances Kemp was in the vicinity much earlier trading in 1838 from Cain Place where Royal College Street connects with Kentish Town Road. A few years on butcher William Pain from Poole in Dorset had settled in Cain Place at number eleven with his wife and three children. And in Monte Video Place, that fronted the opposite side of this triangle, we see butcher William Harris living with his young family at this period.

In some quarters there existed the widely held and mistaken belief that all shopkeepers were successful and wealthy. An example to the contrary and one of many is Edward William Dober, initially a journeyman cheesemonger from Tottenham Court Road, then shopkeeper of Providence Row, Kentish Town; who traded as pork butcher and eating house keeper in this road, before changing to licensed beer retailer until finally being sued and committed for debt. In many instances shops developed from stallholders, paying the property owner for the privilege of setting up in front of their house; a practice still evident in later years as shown by Price Bros meat stall, positioned in the small forecourt in front of 260 Kentish Town Road.

The stall was situated between Osmond's dinning rooms and Atilio Maffla's Central café and restaurant adjacent to the Oxford Tavern. The sale of meat from this location formalised when George Frederick Price dispensed with his stall and opened a butchers shop at this address. On the first storey windows above maybe noted the caged birds, a popular pastime from the late 1890's. Usually a linnet type bird attributed to the unproven derivation of the term Cockney from cock linnet, one of several fanciful theories (misshapen chicken egg) which still abound. A mention here of the Butchers-Bird a migratory red-backed shrike that breeds in England, so called for the habit of dissecting his prey like joints of meat on sharp thorns.

Less obvious to the passer-by, but of immense importance to tradesmen was the opportunity afforded by the provision of rooms underneath or large cellar, this latter area often extending underneath the pavement and originally used for storing coal. The butchers adapted these areas to suit a variety of needs by installing slate salting beds, ham and bacon smoking sections or converting the rooms for meat or small-goods preparation. Many became cold storage areas using ice and later installing mechanical refrigeration, and as was often the case, in preference to using the stairs the meat was lowered and raised through a trap door in the shop by the use of rope pulleys. It is also noticeable from records how much basement workplaces were favoured by the bakers' and cobblers' trades who for reasons of scarcity often shared this type of property with other trades. In 1841, the number of individuals (males above 20 years old) engaged in the trade of boot and shoe makers and menders in the metropolis of London totalled 22,400.

Alterations to buildings to accommodate shops were not all carried out piece meal. A complete row of private houses owned by one landlord, erected in 1837 west of the Eagle public house in Camden Road towards the canal bridge, shops were incorporated within the building in 1853; and permission was granted by the Metropolitan Board of Works in 1864 to build shops on the forecourts of houses on the west side of Hampstead Road. The extent that butchers were themselves property owners has not been fully explored but evidence in the following example verifies this did not guarantee immunity from financial failure.

On the Eagle Tavern side fronting Great College Street, later prefixed 'Royal', were the premises of George, John Henry and Alexander Augustus Richards for brevity styled under the title Richard Brothers; an outwardly successful meat retailing business with branches in Hackney and Holloway. In February 1873, however this family of meat retailers and property owners were adjudged bankrupt. The reasons are unknown, nevertheless on the 8th day of July a general meeting of creditors held at the Royal Court of Bankruptcy, Basinghall Street, trustees were appointed and dividends to the amount of one shilling and five pence half penny in the pound were agreed; this was a protracted affair involving property that lasted several years until 1877 when the bankruptcy order was closed. We also read of William Jepson alias Jesson, citizen and butcher of London, with property connections to Kentish Town.

In the records of the Prerogative Court of Canterbury there survives the will of Kentish Town butcher Phillip Arrowsmith dated 25th February 1813, and probate papers from a later date in 1889 relate to Kentish Town resident and butcher Joseph Rippington who traded outside the area at 24 Upper Marylebone Road. He later lived in some style with his wife Hannah in Chielevey,

Berkshire having from choice or ill health retired at the age of fifty-five. Of one meat trader we have no doubt, butcher and builder William Cook of Somers Town. His shop has been recorded in 1805 at no 1 Southampton Place, Camden Town, after securing a lease on ground from the Dean and Chapter of St Paul's at five shillings per year. He built a row of seventeen cottages, Cooks Terrace on the edge of Agar Town, later the site of the Workhouse Infirmary in Pancras Road; the opening ceremony in 1885, attended by no lesser personages than Lord Mayor Nottage and the Sheriffs of London.

By 1861 the official census population for Kentish Town recorded 44,317 inhabitants, the establishment of butchers shops and ancillary trades increasing in tandem with the various building schemes in progress. Meanwhile, the intrusion of railways schemes continued to despoil the area despite vehement objections from local inhabitants which were usually overruled; the railway companies citing the cause of improved infrastructure to bolster the habitual use of compulsory purchase orders. In November 1861 butcher Edward Davies, 7 Great Green Street, later 77 Highgate Road, was served notice that the Midland Railway Company intended the compulsory purchase of the plot of ground on which his shop and premises stood; his displacement one of concern to other family members consisting of wife Harriet, step daughters Emily, Harriet and Jane in company with Susan a female servant and three butchers men all living above the shop. The details of the outcome are unknown, but St Pancras Vestry minutes for December 1872 reveal he was still in business and granted a slaughterhouse licence for premises numbered 137 Highgate road. The name Great Green Street was often used for the entire length of Kentish Town and Highgate Road. The Highgate Road section frequently referred to as Green Street when named separately with or without the abbreviation U. K T. viz Upper Kentish Town.

The main thoroughfare, Kentish Town Road that extends proper from Camden Town tube station in the south to the junction with Highgate and Fortess Roads in the north, as to be expected attracted the highest number of shops of all kinds. The road has traditionally provided a boundary between contrasting buildings and planning styles, to the west high density and to the east relatively spacious layouts; a significant factor in determining the location of butchers shops, the majority of which opened in west Kentish Town. Such a huge influx of people and businesses meant Kentish Town Road, that had followed the line of natural contours the early travellers had found to their advantage, was itself in need of adjustments; and during the 1880s schemes requiring nothing more sophisticated than a pick and shovel to relieve the bottleneck where Kentish Town Road and Royal College Street converge and again to widen a section of road opposite Prince of Wales Road. The narrow pathway that led to Anglers Lane was opened up when the vestry purchased and demolished 229 Kentish Town Road, to widen the entry at this

Established 1842.
By appointment to H.M. the Queen By Appointment to H.R.H. the Prince of Wales

FAMILY BUTCHER
G. W. GRANTHAM
75 and 76 Park Street & 157 High Street, Camden Town.

G. W. Grantham begs respectfully to announce that the alterations and additions to his premises in Park Street are now complete and every facility is afforded for a largely extended trade.

To celebrate the opening of the new premise and with a view to encouraging the wholesome tendency of the present day to ready money transactions. G.W. Grantham has determined upon making a considerable reduction on all cash orders and weekly accounts.

Customers may rely upon being supplied with First-class Scotch Beef and South Down Mutton. The favour of a visit of inspection is respectfully solicited.

Price lists forwarded on application.

Figure 6: G. W. Grantham, Camden, Town, 1884.

point. Here and there antiquated buildings protruded awkwardly giving a disjointed appearance to the shop parades, but on balance the local retail sector was in good shape to meet the demands of a main shopping area. In the year 1912, Kentish Town Road accommodated two hundred and forty-one individual and multiple retailers offering a myriad of goods and services. Indeed, it might be said that the number of weather blinds down on a rainy day almost provided one continuous umbrella for the entire length of the road. The period between nineteen hundred and the outbreak of the First World War being undoubtedly the hey-day for shoppers and shop owners alike.

ANCIENT TO MODERN

An interesting parallel between butchers, fishmongers, greengrocers especially single shop owners, was a reluctance to update equipment and interior decoration. These three trades were notorious for making do, in many cases not for want of financial resources but in the self-defeating misconception, meat, fish and vegetables were incompatible with modern merchandising. The innovators, J H. Dewhurst and Mac-Fisheries, would lead the way in the nineteen twenties; meanwhile the likes of butcher G. F. Kimber, one of many examples remained unchanged until the end. The spectre of antiquated shops alongside modern grocery multiples and drapery stores Herbert Beddal, C. & A. Daniels and others in the High Street was as disagreeable as it was illogical. The dilapidated condition of some local shops aroused the disapproval of John Betjeman, (1900-1984) born in Lissenden Mansions, Kentish Town, who had a waspish thing or two to say about 'squalid shops' in Highgate Road. His architectural temperament more in tune with the splendid decorative Tudor studded door he spied at the entrance to a butcher's shop whilst on his travels through Somerset. Such was his astonishment he wondered if there was another butchers shop 'in all England' that had such a grand entrance. The shop was owned by Mr D. Woods in the village of Axbridge; no stranger to the meat trade himself, Betjeman lived for a time at 43 Cloth Fair, located between the Butchers Hall, home of the Worshipful Company of Butchers and Smithfield Market. He is buried at St Enodoc's Church, Trebetherick, Cornwall.

The full scope of refurbishment to the premises in Park Street announced by George Grantham is impossible to say other than ten days after 15th March 1884, the date this advertisement appeared in the Camden & St Pancras Gazette, he signed a lease for a term of forty-two years at a yearly rent of £160 pounds for both premises. There were good reasons to modernise, not least the benefit to existing customers and hopefully attracting new ones, shown in increased sales. Of note, before the turn of the 19th century, Park Street was often designated as being in the boundaries of Regent Park area.

First noted in Camden Town in 1842 George William Grantham served as St Pancras vestryman and was in a substantial way of business. In 1892-3 he acquired leases for numbers 155, 157 and 89 for terms of fifty, twenty-one and seven years respectively for butcher's shops in the high street. Already a wealthy man, he lived above his Park Street shop and his personal finances considerably increased later with his involvement with Lidstone Butchers Ltd.

There had been, from the latter half of the 19th Century, a swath of innovative ideas and general improvement in shop fittings and construction, spawning a local industry of shop fitters, sign writers and weather blind makers. For example Cattell & Son, 63 Fortess Road, Kentish Town manufacturer of spring and roller shop blinds. Most important were the local construction builders

Figure 7: S. Simons, 216 Kentish Town Road, 1903.

like W. Graham & Co who advertised as 'Shop Front Builders' by cognizance specialising in this type of work. Today modern fascias and other cosmetic features conceal their hidden history, but something of the original owner's intent can be seen in old photographs; most rewarding the decorative trusses and pilasters, the rectangular column that fronts the party wall.

This is especially true of Thomas Burkett's shop; however by this period hard wearing easy clean marble replacing ornate plaster moulding that crumbled with time. Additional exterior features were the introduction of fresh air by way of wrought iron grills and bars above the door and window, especially welcome in summer when the shop was closed, although 'fly cemeteries' (sticky fly paper) would remain a feature in food shops for many years to come. The position of weather blinds and awnings in some cases meant repeating the shop name twice, on the wall above and below on the facia, or if this was impractical printed on the main and side blind. Local authority regulations required weather blind supporting arms to have 7 feet 6 inches clearance from the pavement, for extra safety and abiding by the law many were fitted with chains.

Interior improvements like stainless steel replaced hanging rails made from iron, and glazed tiles covered white washed walls, these were initially expensive, but were hardwearing and easy to keep clean. A novel use as it was then advertised for a natural occurring material was slate in use by the 1840s for counters, window beds and shop stall risers. A fragile material, and time consuming to clean, it fell into disuse to be replaced by marble. Intriguingly, Sydney Whiting of 9 Maida Hill Westminster in 1865 had patented "improvements in shop and other counters and surfaces on which money is placed in passing it from one person to another" his invention and outcome unknown.

Modern sectional cutting blocks made from hornbeam, hickory or maple from America, replaced elm planks and oak tree trunks, the latter although hardwearing contained gallic and tannic acids which blackened with time. As early as 1863, local inventor Thomas Powell, 30 Princes Terrace, Regents Park, was working on an improved butchers chopping block. Most noticeably early domestic shop doors gave way to a wide expanse of empty space, as fixed doors presented a psychological barrier to customers. In the shop of Sidney Simons & Co, and disregarding the over ambitious slogan 'Branches Everywhere' directly beneath the facia, they had four branches, we see the innovation of roller shutter blinds and the replacement of gas lighting with electric. The ravages of time have been kinder to some shop fronts than others and many are the circumstances that have influenced the final outcome; yet all began with hopes and dreams that in part were reflected in the design and attention to detail. For the passing public untold snippets of local history lie beneath the modern fascia.

From an earlier period in Kentish Town, gas lighting was a contentious issue for ratepayers who were less than enthusiastic during a shop versus street lighting dispute with local authorities. This prompted one unnamed resident in 1831 to publicly air his grievance in a letter addressed to

Figure 8: Valentine Lunch & Co, 280 Kentish Town Road, March 1903.

the editor of the Times newspaper, in which the writer deplored a situation that had arisen whereby "gas was being laid down to accommodate shopkeepers as choose to light the interior of their shops with gas. Whilst elsewhere in Kentish Town various bad and evil disposed characters under cover of darkness are free to commit robbery". The St Pancras Council were the first municipal authority to adopt the electric light system, upon application in 1892, to the Board of Trade to produce and supply electricity for public and private use. The first public thoroughfare to be lit by electricity within the borough was Tottenham Court Road. However electric lighting in shops, workplaces and houses was not widely available until after the Great War. By then a basic utility service and still not fully implemented post Second World War in many private dwellings.

As time passes the growth of any area inevitably means change and innovation and Montague Place, home to Kentish Town panorama artist James King, south of the Assembly House pub was demolished and replaced by the Midland Railway Station. A similar fate on behalf of the travelling public awaited the adjoining row of houses named Inwood Place with the construction of the underground railway and Kentish Town tube station opened in 1907. The buildings before this event were converted to high street shops, one occupied in 1900 by Valentine Lunch & Co. This company was un-incorporated therefore no official records exist so there is considerable mystery as to the identity of those involved with Mr Valentine Lunch, although a John Page is listed as rate payer. However bankruptcy adjudication records in May 1902, reveal the finances of Lunch & Co also carrying on business at 81 Stoke Newington Road, were in disarray. In August 1903 less than five months after this photograph was taken, orders made on application for discharge declared were suspended for two years. By which time Valentine had found employment as butcher's manager.

The dejected looking thirty nine year old fishmonger is Henry Smith that with his father William traded as W. Smith & Son, later relocated to Fortess Road. In 1905 the Charrington Cross and Hampstead Railway Company acquired this and adjacent properties, Lunch & Co among them after securing a compulsory purchase order and promptly granted advertising rights to Partington & Co for newly erected hoardings whilst building work for Kentish Town underground station was in progress.

SERVICE SECURES SUCCESS

Life in early Edwardian Kentish Town, despite the underlying poverty here and there, reflected the nation's air of stability and routine; each household regulated by the calendar and clock. From Monday through to Sunday, meals, shopping, domestic chores and even religion were each given their allotted space and time.

Figure 9: C. Dray 172 Kentish Town Road, 1904.

Figure 10: M.L. W. Redhead, 172 Kentish Town Road, 1904.

The daily shoppers for good reasons visited the bakers first to ensure bread was fresh from the oven; then in descending order of food deterioration the greengrocer, dairy and finally the butcher, fishmonger, and poulterer depending on the vagaries of the weather; casual window shopping was left to the afternoons. Three quarters of businesses in Kentish Town Road and all but one of the ten butchers' shops there in 1903 were family owned, personal service and attention to detail were the hallmarks. A time when most shop assistants, traders and craftsmen knew their business, discussed their customers requirements and made time to engage in cordial conversation. A time when people dressed-up to visit the shops and the children's nursery rhyme, the butcher, the baker and candlestick maker still rang true in the high streets.

A treat for lovers of shopping in times past is the photograph taken of butcher Charles Dray's shop, which illustrates perfectly the calm of a quite days trading in Kentish Town Road. He opened for business in 1902, although he initially lived away from the shop in a small terraced house with wife Nellie and three children and shared the property in Sandringham Road, Willesden with elder brother George and family.

The previous shop proprietor M.L.W. Redhead there in 1899, and at 233 High Road, Kilburn and of whom little is known, other than he employed fifty year old butcher's manager Edwin Churchill and wife Clara living in tied accommodation above the shop. His middle forename Mathew Lee Winter, with American connotations, giving credence to cursory evidence he was related to American born citizen Daniel H. Redhead a meat salesman in business on Smithfield Market. First recorded in 1878 there followed a series of partnerships, the last with W. A. Darrington, trading in meat casings on Smithfield Market. If substantiated, the dissolution of this partnership in 1898, may suggest financial difficulties leading to a rapid collapse of their retail interests.

On Sundays the road returned to some resemblance of village life past, although the occupants living above the Dray shop made certain as like wise others in the road of their own private oasis of tranquillity by enjoying a small patio garden and conservatory on the flat roof space. A world away from the rough and chaotic atmosphere of Queens Crescent market where discriminating shoppers never penetrated for fear of being seen by their neighbours. It has been claimed by some observers there was snootiness in some business circles to ensure that shopkeepers in Kentish Town Road remained aloof from cheap jacks and vulgar street traders; sandwiched as it was between the lofty social and geographical heights of Hampstead, Highgate and the dynamic materialism of Camden Town.

It was true the vestry were continually sending warning notices to kerbside traders for causing a nuisance in Kentish Town High Road, however this was commonplace to all local authorities throughout the Metropolis and nearly always at the behest of rate paying shopkeepers justifiably concerned over lost sales, many of whom served on the local vestries. How successful their n.i.m.b.y. (not in my back yard) stance is hard to gage, for in May 1912 estate agents Salter Rex & Co, established in 1854, raised an intriguing point when they advertised for sale by auction a house and shop at 198 Kentish Town Road. 'In the centre of a market previously let on a repairing lease to well-known firm of boot manufactures'. The unnamed boot makers were John Kavanagh Ltd, adjacent to butcher John Edwards. The latter named, sited in a parade of eleven shops on the eastside of Kentish Town Road between Patshull Road and Gaisford Street.

The reference to a market may have just been advertising hype, for only a few maverick costermongers still plied for trade in the road; then again perhaps Islip Street where fruit and vegetable stalls were pitched and shared the road with the Hackney cab stand. The term hack a derivation from the French 'haquenee' meaning worn out horse. Interestingly both the premises of Edwards and Kavanagh were eventually acquired by John Sainsbury until in December 1955, relocating to their new supermarket further along the high street.

Figure 11: J. Edwards, 196 Kentish Town Road, 1903.

As yet unmentioned, but of equal importance, were the side or back street butchers shops, that in their day added considerably to the overall numbers of retail butchers in Kentish Town. They flourished not surprisingly, but not all, in the poorer districts through their ability to provide a life line of weekly credit; the service given various colloquial names such as tick, slate and strap, and shopkeepers as businessmen adding a penny or two for their trouble. In the matter of credit, there were two distinct categories, one already mentioned and that extended to those customers who budgeted their domestic finances monthly or quarterly.

The side street butcher, far from being the under dog of meat retailing the old adage 'Many a pig has been sold on the way to market', should be borne in mind, as aside from their immediate convenience to residents the street or road often funnelled human traffic to and from their place of work and other activities; thereby creating considerable opportunities for passing trade.

At random, a few of the many west Kentish Town butchers shops which met this criteria, in no order of occupancy are Cato, Corbett, Bayler, Burkett in Carlton Road, Barrett, Bishop, Searle and Pyle in Castlehaven Road, Jakings, Anwyl, Folks in Crogsland Road, Bailey and Mundy in Rhyl Street, and Gate, Woods and Roberts, Warden Road, Elvidge, Hersant in Harmood Street.

Of these Richard A. Bayler at 78 Carlton Road, a section of Grafton Road until 1937 then renamed as a single identity with Grafton Road. In a small way of business in the area c1900 where he was born, Richard was assisted by his wife Elizabeth and their son William who suffered from epilepsy. The trade was insufficient to provided work for their two other sons and a daughter employed elsewhere. The shop was sited adjacent to the Mamelon Tower public house, then at number seventy-six in this road, and throughout its history more closely associated with Queens Crescent market. Originally the Manchester Tavern the hostelry changed its title in deference to the British soldiers in 1855 during the Crimean War that with the French captured a fortified position from the Russians outside Sevastopol

From an earlier date the Hersant family were prominent in the meat trade spanning several generations of male descendants, most entering and expanding the family business. One of whom master butcher Milton Hersant setting – up shop and licensed slaughterhouse at 3 Prince of Wales Crescent and Harmood Street with more to follow trading as Milton Hersant & Sons in Marylebone and Harlesden; the latter shop in 1940 still in business. Milton lived above the Prince of Wales shop premises with his wife, four sons and two daughters; Thomas the eldest running the Marylebone shop with a Henry E. Hersant until retirement and living at 20 Woodsome Road, Highgate.

In 1856 Milton Hersant serving in his shop at number 47 Harmood Street, became the victim with other local retail businesses of deception and forgery concerning money orders allegedly perpetrated by William Attwell, Of whom it was said repeated the alleged offence in the Prince of Wales public house a present day survivor amongst the ever declining number of early Taverns, directly opposite the Hersant shop The landlady at this time Mrs Suter with employee potman William Golding, giving evidence together with Mr Hersant at the trial. However despite the evidence of Ransom's bank in Pal Mall, locksmiths Bramah & Co, The Sportsman public house, City Road and a pawnbroker in High Street, Camden Town. The Recorder was of the opinion the evidence was to slight to go to jury and subsequently the charges against William Attwell were dismissed.

On the eastern side of Kentish Town: Richard Selway, Thomas, Corrigan, Jackson Bros in Chetwynd Road, John Plaistow in Islip Street, S. Webber, Swains Lane and Williams, Bennett, Atwell, Salt, O'Hara in Leighton Road, and John Plaistow in Islip Street, c1880. There were in addition to high street, market and side street butchers, those sited among other retailers in short parades along important transport routes or secondary connecting thoroughfares; for example Chalk Farm Road, Ferdinand Street, Mansfield Road, Southampton Road and others.

MEATHIST MINIATURE

The butchers shop implements including balance scales once a familiar sight to the inhabitants of the large Roman settlements of London, Colchester, St Albans and Chester would still be recognizable in our society today. The emergence of town and village shops in the 18th century began in response to economic growth and in consequence consumer demand. The oldest surviving retail butcher's shop name in Kentish Town was Hales trading from various locations within the district from c1766 to 1954.

2

THE GUILD CONNECTION

Central to all trade activities were the Guilds and City Companies, a group of individual associations of many religious origins, which emerged during the medieval period to promote and protect their own craft or calling. Of these the merchant guilds, or twelve great companies as they became known, were the most rich and powerful; wielding great influence in the affairs of the City of London; establishing a system of local government the basis of which still exists in the city today. Within and in addition were those craft guilds that were essentially hands-on and by definition service trades, particularly the victualling guilds, such as Butchers, Bakers, Fishmongers and Poulterers. The true origins of the guilds or companies are disputed; some historians believe similar fraternities evolved in Anglo-Saxon times, for others not until the middle ages; however it is from this latter period most records survive and from where it is appropriate to begin.

The butchers guild, or Worshipful Company of Butchers as they later became, are among one of the oldest guilds with a long and eventful history. The origins of the guild are believed to derive from the butchers of Eastcheap flesh market disgruntled with constant complaints and subjected to punitive controls and regulations imposed upon them by the Crown, they formed a fellowship or fraternity. First official mention of a London butchers guild occurs in the Pipe Rolls of 1179-80, one of eighteen guilds listed as owing money to the Royal Exchequer and being adulterine, that is operating without authority of the King. The widely held view is the naming of the adulterine guilds was probably not financially motivated but an attempt by the crown to control the guilds lest they became too powerful. A not unexpected reaction, for in feudal times from a modern day perspective these guild fraternities must have seemed like militant trade unions, a fundamental right denied to all workers until 1823.

The butchers, of whom it must be acknowledged, were in the unenviable position of seeking recognition of their mystery while at the same time engaged in a trade the nature of which affected their standing with both public and state. Nevertheless, a tolerable relationship prevailed between the butchers and higher authority, their status immeasurably enhanced following the grant of a Royal Charter by James 1 under the Great Seal on the 16th September 1605. However for the inhabitants of the city, and the entire country who were caused 'great nuisance' in the manner the trade was conducted it remained a public relations disaster which did not improve until slaughtering, the most unpleasant aspect of the trade, was banned from highly populated areas. Vested with various ordinances and powers by successive Crown and civic authorities of the day, together with their own domestic regulatory court, the guild tried to protect and control the art or mystery of their profession within the City and two miles beyond, including other towns and cities where the guild existed. Their sphere of influence and power covered markets, hours of business, prices, restriction of sales and slaughtering during religious days, including non-freemen and hawkers. Of paramount interest were matters of health, meat should be wholesome "good for mans body" even ancillary trades tanners, curriers and chandlers all came under the watchful eye of the guild. Judgements from the earliest court records show rules were strictly enforced.

The guild jealously guarded its craft and only freemen, that is guild freemen not Freemen of the City of London, were allowed to sell meat or open a butchers shop on completion of an indentured apprenticeship to a master butcher; this was normally of seven years duration although twelve years or more was not unknown, followed by one year as journeyman. Entry into the craft was by way of letter or personal application and later newspaper advertisements, the London butchers apprenticeship being the most prized. Although inexplicably, applications from young men and boys domicile in London were few in number as were young women from the early lists. Yet apprenticeships it was said were always open to both male and females, however the guild records would seem to suggest otherwise as the masters list 1605-2004 is by definition male dominated. Conditions of butchers' apprenticeships in medieval times were comprehensive and covered both health and morals. Masters would provide full board and clothing and in return expected good behaviour, household chores and not to marry without permission, and for good measure suffer correction as his master shall deem appropriate.

The sounding of the Bow Bells of St Mary le Bow church, associated with the term cockney, called time on the frolics of the young apprentices each evening. Those needing more persuading met with the Bulbegger, a 16th century local enforcer, the original Bogeyman and used by generations of parents since to coerce disobedient children. The actual butchery skills were taught by demonstration and practice as few apprentices had scholarly backgrounds. The years of service for indentured apprenticeships was universal and strictly controlled, a term of not less than seven years training was made obligatory throughout the country by the Statute of Apprentices of 1563; under the same Act any boy who became apprenticed without his parent's permission could not be released unless the master consented. Any serious misbehaviour was quickly bought to the attention of the worshipful company, and many young lads found absence from home and the rigours of the trade distressing and left their employment, a course of action that immediately put them on par with the common criminal. As the following advertisement which appeared in the Times Newspaper, 26th May 1846, aptly demonstrates.

> *"TWO POUNDS REWARD"*
>
> *Absconded on Friday night, March 13th, from the services of their master, Mr Ell, butcher, of Clifton, Beds. Two apprentices named William Branson and Joseph King, the former is about five feet five inches in height has dark brown hair and thick lips and was dressed in a butchers frock. The other apprentice is about five feet seven inches in height and when left his master was dressed in a butchers frock, corded trousers and high shoes. Whoever will give information of the said persons, shall receive the above award and all reasonable expenses".*

The bulk of time served, apprentices it seems chose not to become freemen, instead returning to their home village or town with their parchment certificate to practice their trade. There were numerous cases of non-freemen continually trying to circumvent the system, which in reality was a closed shop regime. Astonishingly this archaic piece of legislation, first introduced in 1275, was not revoked until 1856 unfortunately too late for Messrs Green and Hydrons the last to be so summoned before the court in March 1851 for trading with out being free. This was a last desperate attempt by the guild to enforce the unenforceable, for by this late date London had long since outgrown the city limits with a population fast approaching two and a half million. Unwelcome as it might be for the butchers company the stranglehold had to be broken, for at this period there were no fewer than eighty-two other London Livery Companies from Tin Plate Workers to Tobacco Pipe Makers all capable of monopolising their trade; and many of the guild regulations were seen by the government as restrictions of trade and being revoked or taken over by none partisan authorities. Butchers throughout the country not subject to the jurisdiction of the guild frequently set-up shop where they pleased, effectively removing themselves from any guild control; a process that had been taking place since the 1680s with the emergence of the provincial butchers shops during the so called commercial revolution. Census figures collated for occupations in St Pancras in 1851 reveal there were 467 butchers shops from a population of 166,956, and it is from the surviving apprenticeship registers of binding at the Guildhall we see that process in our own locale.

Of particular relevance is the original parchment document of apprenticeship indenture from which we learn on the 6th January 1674-5, Charles the son of George Sanders from Highgate, was apprenticed for seven years to future master of the company, butcher George Clipson of London. Despite searches, other than the family surname and Sanders has been entered as Saunders in the register, there is no record of Charles being admitted as freeman. This was a common occurrence and could be for one of several reasons; an official discharge following legal annulment of the indenture, the apprentice absconded, and could not be apprehended. The majority worked as paid journeyman on completion of the apprenticeship whilst others returned to the family business or opened shop on their own account. We cannot be sure, but George Sanders may have owned a shop in Highgate to which Charles returned. From the same district a one line entry in the apprentice book reads "Nick Bennett son of Thomas Bennett 1728 Highgate", and a similar curt reference undated relating to James Reynolds Highgate, summoned for quarterage i.e. payment of dues. One young man who did progress was Thomas Hathaway, son of Charles Hathaway

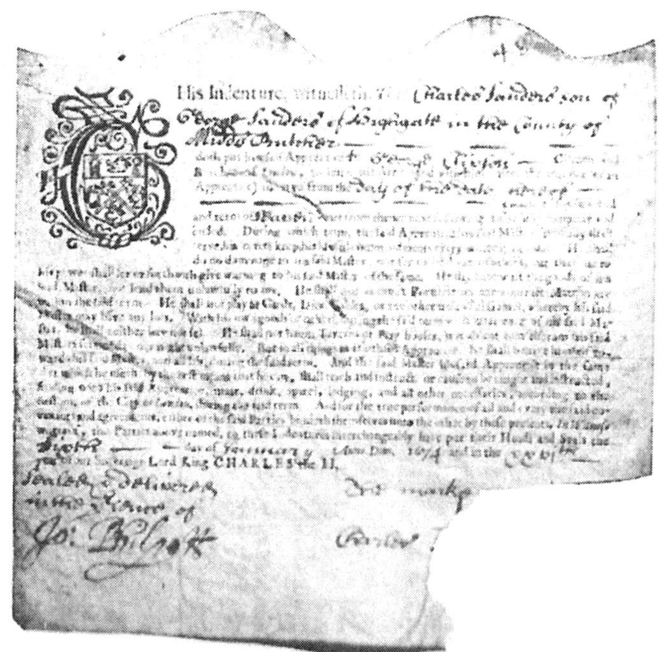

Figure 12: Indenture of Apprenticeship, 6 January 1674-5.

from the parish of St Pancras, the duration of his apprenticeship seven years ending in 1763 thereafter made freeman on 3rd June in the same year.

Apprentices were left in no doubt what was in store with the publication in 1747 of 'The London Tradesmen', a summary of employment opportunities and advice for parents and youths. The author R. Campbell has some scathing remarks for quite a few trades and professions whom he believed offered little future prospects and even less wages.

> *'The butchers generally require more skill to learn their trade than any other of the victualling branches we have mentioned. They must not only know how to kill, cut up, and dress their meat to advantage, but how to buy a bullock, sheep, or calf, standing: They must judge of his weight and fatness by the eye; and without long experience are often liable to be deceived in both. Butchers are necessary; yet it is almost the last trade I would choose to bind a lad to. It requires great strength and a disposition no ways inclinable to the coward. A lad may be bound about fourteen or fifteen. The wages of a journeyman is not much more considerable than that of a common labourer'.*

The book did much to alert parents of the pitfalls awaiting their son or daughter apprenticed to the wrong trade or master, but conditions of employment for young apprentices continued to deteriorate. As public anxiety grew it became essential to protect them from exploitation by law, and in 1802 Robert Peel, against considerable opposition, pushed through the Apprentice Health and Morals Act. That said there was no denying those engaged in the practical side of the butcher's trade stood low in the social class.

It is almost obligatory to mention journalist Daniel Defoe, originator of the first real newspaper The Review est. 1704, and author of the novel Robinson Crusoe. Of interest here; he mentions the retail butchers in Highgate in correspondence during the writing of his three volume travel tour books Through the Whole Island of Great Britain' published between 1724 and 1727.

> *"The Jews have particularly fixt upon this town for their country retreats, and some of them are very wealthy; they live there in good figure, and have several trades particularly depending upon them, and especially, butchers of their own to supply them with provisions kill'd their own way."*

The butchers trade with which Daniel had more than a passing acquaintance as his father James Foe, was a butcher in Cripplegate, London and elder statesman of the company. Although unlike his father, when elected freeman of the Guild in 1688, paid a fine of £10.5s and was relieved of all obligations to serve in office, preferring to use his talent for expressive writing that he undoubtedly possessed and choosing the pen name Defoe. By coincidence Daniel's last resting place was alongside John Buyan and William Blake, who mentions Kentish Town in one of his poems, is a stones throw away from Smithfield Market, for they are buried in Bunhill Fields, a corruption of Bone Hill, a prehistoric burial site.

Away from the guild women are frequently recorded as shop proprietors, the mantle of responsibility often thrust upon them by unforeseen circumstances. In most cases the husband absent through incapacity, death or call to arms during a national emergency. In Mrs Hannah

Figure's case, trading from No 5 Great Green Street, Kentish Town, between 1841 and 1852 the reason is unknown, however with the responsibility of daughters Alice and Harriet the presences of journeyman butcher Edward Dorrmer living on the premises would suggest a division of the practical and financial aspects of the business. Nevertheless women have through the centuries proved they are more than capable of working in trades formally considered exclusively men's work. Other examples are Catharine Knight, Jane Good, Charlotte Kirkland, Martha Kempton, Dorothy Nickles in Kentish Town, Elizabeth Attkins, Lee Harper in Highgate and Mrs M. Hannah in Hampstead.

Of particular importance and fascination, the origins of general trade names still commonly used today; many of them post 1066 and the battle of Hastings, a date involuntary etched on every school pupils mind. The arrival of the Normans with their ready made aristocracy and legal system began a two tier feudal system of privilege and rank, separated by two languages English and French. The Norman lords, as the dominant authority, selfishly procured the prime cuts of meat and applied their own French names; for example boeuf, mouton and porc, thus evolved beef, mutton and pork as applied to the choicest cuts, as in beef or pork steak and leg of mutton. The Anglo Saxons as the underlings used the Old English terms for ox, sheep and pig that is oxa or bule, sceap and pigge for the inferior or less palatable parts of the animals eaten by them, as in ox or bullocks, sheep and pigs' heart and liver. It seems the Normans bested the English again, as the derivation of the word 'Butcher' is from the old French '*bouchier*', which in turn is from the Latin '*carnifex*'.

BUSINESS AND PLEASURE

In 1774 sufficient guild members were domiciled or acquainted with Highgate and Hampstead, to justify holding what had become known as the 'anniversary dinner' at the Bull and Bush public house, Hampstead. The tavern forever associated with the song 'Down at the Old Bull and Bush', which was immortalised by Australian born music hall artist Miss Florrie Ford who also lived temporarily in Hampstead. The reason for the event was to celebrate the judgement given on 23rd June 1762 upholding the legality of the Act of Common Council requiring all butchers in the city to be admitted into the freedom of the Company of Butchers. It was decided to mark the occasion by holding a dinner each year on a date nearest that day. The outings were originally held in the city, but it was later agreed "that a drive into the countryside by coach and four would be most enjoyable", and thereafter visits to venues further afield were undertaken.

These particular junkets came to an end in 1789, but as London grew there would be many such gatherings of meat trade organisations in the years ahead. The emergence of local associations and national federations in the trade saw many activities, previously the prerogative of the Worshipful Company of Butchers, transferred to them. It would be impossible to explain them all in detail here, but official meetings and social events were a characteristic common to all the meat trade organisations. The variety and scope evident to name just two; the 'Chester Journeymen Butchers Provident Association' and the 'Jewish Butcher's Association' formed in the early eighteen hundreds.

Of special interest, the third annual reunion and dinner associated with St Pancras Section of the Union of London Meat Traders held on the evening of Monday 16th February 1924, at the

Figure 13: Kentish Town Fields towards Highgate, engraving, J. Storer, 1805.

Midland Grand Hotel, Euston Road. The gathering ostensibly a social occasion, after the toast of the 'King', the master butchers among them relished the opportunity to air their grievances on the various government policies that hindered their livelihood. The local Kentish Town proprietors in attendance included Arthur Coggan, George Kimber, Harry Tipple, Robert Warren and George Wheeler. Their concerns on matters of working hours, meat control, shortage of meat and competition from multiple shops were recurring problems for numerous generations of butchers.

It was said the entertainment was varied and professional, amongst the company of artists Miss Marion Ruth and Fred Rome in cameo conversation, with appearances by Miss Dorothy George and Messrs Hill, Covell and Stoud. The music provided by the Rosa Orchestra under famous composer Lillian Ray. Though they are rather less prominent today local associations within the National Federation of Meat and Food Traders continue to meet for official business and social functions. In addition to national and regional trade associations, many local businessmen have served their community through their involvement with local councils and the Chamber of Commerce. The vestry records in particular show a preponderance of businessmen, including many butchers shop proprietors. Whilst there is no suggestion of any impropriety there were conflicts of interest, most markedly during the long running battle to abolish private slaughterhouses within residential areas.

From the latter half of the 19th Century the area around the Metropolitan Cattle Market, opened in June 1855, became fashionable for Guild members to live and in particular Camden Road; the proximity to the new market influencing their decision; twelve guild members choosing this area like Benjamin Venables and his sons Edward and Arthur, and John Collins, Smithfield meat wholesalers and cattle salesmen. Future guild masters Benjamin and John were neighbours residing at Athol House and Argyll House respectively at Camden Villas on the north side. Edwin Seymour Lardner living in Hartham Road could almost hear the baying of the animals. From the 1881 census, cattle salesmen resident in Camden Square included fifty-five year old Henry Trotter living at 152 Agar Grove.

At number ninety-nine in the same road salesman Mitchell King, with fellow border and cattle dealer James Corcoran, with market cattle drivers Thomas Mcmanage and Michael Smith constituted a separate household in rooms on the upper floors. At number 3 Camden Terrace lived seven members of the Merten family from Germany, head of the household George and his brother Erust in partnership as cattle salesmen. Of further interest forty-seven year old George Halfield at 11 St Pauls Crescent, a Cheshire man employed as cattle constable at the Market. Within Kentish Town, Liveryman Edwin Cox, son of former guild master Thomas Cox, was resident at 11 Rochester Terrace from 1876 until his death on 1st July 1884. At this period, there was hardly a road within a two mile radius of the market that did not have a resident with business or employment at the market.

INSIDER DEALING

Further proof of activity of a more devious nature is provided by a record of a meeting of the Master Butchers under the chairmanship of Thomas Dalby, convened at the Nags Head Tavern, Leadenhall Street on the 1st August 1789; which heard evidence that carcass butchers and jobbers regularly meet in the fields of Kentish Town, Mile End, and Knightsbridge to purchase cattle and thereby manipulate prices - forestalling, (to purchase goods in advance of the market) on Smithfield market. The names of the fields in Kentish Town are not recorded, we can only speculate they may be among those surviving as such, or those from which today's names of council apartment blocks have been derived for example: Barns Close, The Forties, Landleys Field, Long Meadow, Tanhouse Field and Mutton Place. The last named Mutton Place situated in Harmood Street nearby the once 'Shoulder of Mutton Field', a field covering eight acres triangular in shape resembling the scapula bone, which undoubtedly accounts for its name. It is known John and Mary Harmood were tenants of the Tottenhall Court Manor c1800 and occupied a field there, possibly the one in question. A block of flats 'Tottenhall' in Ferdinand Street is named after the Manor. The influence of the meat trade in Kentish Town is again affirmed with the naming of 'Butchers Field', a 3 acre piece of grazing land roughly on the line of Crogsland Road that bounds the eastern side of Haverstock Hill School. The name place game was a practice with limitless potential, repeated from field to lane to street to river, pond and brook. As for example in Leg of Mutton Pond, Hampstead Heath and ladies fashion, Leg of Mutton Sleeves or "manches a gigot", officialdom has since intervened, and naming of fields for registration purposes is no longer permitted under agriculture regulations.

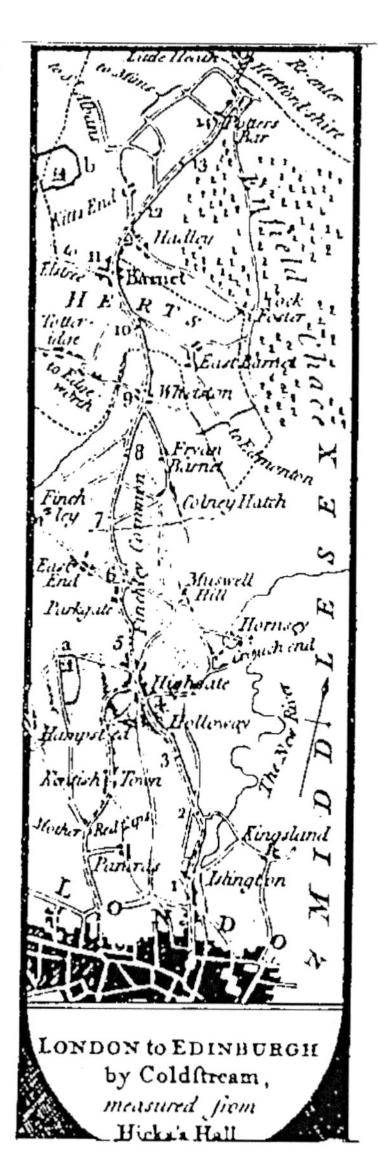

Figure 14: Drovers' Roads, 1785.

In 1840 Smithfield cattle salesman William Vorley residing at 6 Strahan Terrace, Islington secured a one year tenancy agreement for much of the pasture land mentioned, in addition Pear Tree Field, Hill Field and Kentish Town Field; a mutely profitable side of the business for William and his cattle agent banker Chas Hill & Co, of 17 West Smithfield; also John War formally of Caroline Place the previous name of a short section in Chalk Farm Road.

A cattle dealer/salesman and commission agent his place of business and letters the Ram Inn, West Smithfield. He was sued for debt in May 1851 and found temporary accommodation in York Place, Kentish Town Road. Another example is found in cowkeeper and cattle dealer William Candish running Fortess Dairy, Fortess Terrace, and renting several fields between Kentish Town and Highgate Hill, out of business for debt by 1858, living in Shepherd's Bush.

Returning to insider dealing, these forestalling cartels were interceding between the farmers and retailers to alter the natural levels of supply and demand, creating shortages and increased prices for retailers and customers alike. So serious was this matter considered by the retailers and others, concern was expressed in the House of Lords; however a Bill to stamp out the practice was unsuccessful. One estimate puts the total number of animals involved from all three areas already mentioned at five hundred per day. The extent may have been exaggerated, but the fact Kentish Town had been named would suggest considerable involvement. At the time of the Nags Head enquiry, over 140,000 cattle and 1,000,000 sheep arrived annually at Smithfield market, nearly all travelling the drover's stock roads converging on London.

Of the three most widely used routes, one along the Great North Road passed through Barnet then through Highgate and on via Holloway Road, Islington, and finally to Smithfield. However the drovers route of entry into Smithfield, on one particular map taken from Paterson's British Itinerary 1785, also clearly shows an alternative route from Highgate via Hampstead, Kentish Town and through Camden Town passing the Mother Red Cap. The Hicks Hall in St John Street, close to the City boundary was the starting point for measuring distances on the North Road, and in its time the Sessions House for Middlesex.

It was Highgate's strategically placed location that afforded the opportunity to divert animals to Kentish Town for dishonest deals to be arranged. For Kentish Town was then a provider of grazing land for the purpose of resting and fattening the animals before being taken to market. In addition being one of many agricultural larders in the surrounding hinterland supplying the inhabitants of London. Where corn, hay and straw dealers came to bid and replenish their storehouses from the likes of Mr Prickett with 2 Lots for sale; a growing crop of wheat and rick of hay standing on two fields by the side of the high road, leading to Millfield Farm, Kentish Town Hill. The lane leading to the farm the given name of present day Millfield Lane on the eastern side of Highgate West Hill, alias Kentish Town Hill.

There were three classes of drover, the country drover who on arriving on the outskirts of the city handed his charges over to his London counterpart, the Town drover for the final journey to market and the Smithfield drover hired by the butchers to herd the animals, once purchased, to the slaughterhouses. The country drover in reality was neither a shepherd or herdsman, but a hard headed licensed business negotiator, highly skilled at transferring large numbers of cattle, sheep and money the length and breadth of these Islands; in addition relating gossip and news of the latest farming innovations. Self interest ensuring their safe arrival in good condition, as often drovers purchased the animals beforehand from the stock farmer.

The drover, as a devil may care flamboyant character with a twinkle in his eye and a flock of sheep, existed only in the imagination of generations of novelists and poets. A myth reinforced by the dandified Smithfield drover, personified in W. H. Pyne's 'Costumes of Great Britain' published in 1805, and perpetuated with folklore of songs and curious customs; of the many to choose from, the custom of "Swearing on the Horns" originally a form of solemn initiation ceremony for Drovers with satanic undertones.

The romanticised image reinforced in 'Hugh the Drover', a romantic ballad opera composed by R. Vaughan William enacted in various parts of the country, the parody of which becoming a night of entertainment in taverns, inns and ale houses along the drove roads. In particular the Highgate area, a favourite overnight resting place before the final journey to Smithfield market. The Gate House Tavern, Highgate, where it is recorded the ceremony dates from 1688, and one of nineteen venues in the Highgate area in 1826 which held a burlesque version of this ancient ceremony.

Of these, which included the Bull and Last in Kentish Town, The Horns, and the Wrestlers Arms on the North road, eleven used stag horns, seven used ram horns and one bullock horns. As far as the publicans were concerned it was of course an excuse to extract money from a gullible and willing public prepared to recite ridiculous oaths in between consuming copious amounts of ale much to the landlords delight. Of peculiar oddity, the boundary of the northern most part of St Pancras and part of Hornsey lay within the Gate House tavern grounds.

Figure 15: Swearing on the Horns, Highgate, c 1906.

The reality was somewhat different with drovers and livestock farmers continually at risk from cattle, sheep and horse rustlers that from medieval times plagued the countryside. Thomas Cuttes the younger and Robert Mytchell of Smithfield Bar appearing at Middlesex sessions in September 1613 to answer questions of stealing twenty four sheep found in a close rented by Mytchell situated in Kentish Town. The examination postponed until the next sessions because of fear of execution against him when the Sheriff find him 2s. 10p.

In the same year yeomen Edward Sturdifall and James Wilson of Kentish Town for stealing three lambs each worth 5 shillings from neighbour George Fowler. Having no bill of sale Edward was sentenced to be hanged, the said James still at large. This category of criminality still evident one hundred and fifty years later when on the 21st of October 1867, Joseph Jones a 16 year old licensed drover was instructed to drive thirteen long wool sheep owned by Richard Lathbury from a fold near the Brecknock Arms in Camden Road to a piece of grassland in Torriano Avenue. By

6 o'clock that evening the sheep had arrived safely in their temporary fold and were enclosed using hurdles, a short while after all the sheep were gone. The missing sheep were noticed later being driven along Kentish Town Road, where two witnesses positively identified Jones as the drover.

Their identification skills confirmed next day when the witnesses picked out Jones from among 150 young drover lads in the Metropolitan Cattle market, Islington. The suspect was committed on a charge of stealing, however due to an oversight the mother of the boy and a lodger in the house were not examined before the committing magistrate and when asked at the trial both confirmed the defendant was at home on the night in question. The jury were unable to agree on a verdict and Jones was discharged.

In the same year on 6th June, at Clerkenwell Court, John Read a cattle drover was charged before Mr Cooke with stealing 32 Isle of Wight lambs, the property of Mr John Fuller, master drover, 3 Lyons Mews, Maida Vale. The stolen sheep part of a flock numbering 71 was marked with green paint on the head and shoulder, penned in his field with hurdles at Tufnell-Park Road. Mr Fuller stated he visited the field which is building ground at 6 o'clock on Monday the 27th May last and saw the lambs safe, when he returned next morning the lambs were gone and the ground being hard he could not trace which way they went. Luckily an expert witness was in the vicinity in drover Richard Allcock of North Road, Highgate, who testified he saw the defendant whom he knew between 10 and 11 that night with a drove of lambs of that particular breed coming down Gloucester Place (Leighton Road) into Kentish Town Road. A host of other witness mainly from the Metropolitan Cattle Market appeared for the prosecution and John Read was remanded in custody having previously been admitted to bail on a charge of stealing two bullocks.

When the railways overcame their initial aversion to transporting animals the country drover's way of life gradually came to an end. How widespread the problem of rustlers had been, and if the visiting drovers or indeed local farmers in Kentish Town were implicated in forestalling, or similar criminal activity at any period is a matter of conjecture, we can be assured however the subject was a topic of conversation amongst the livery members in the confines of Butchers Hall.

SILVERSIDE & GARRETT

Whilst there is no suggestion the next two gentlemen were anything less than honourable, it would be naive to suppose they were blissfully unaware of forestalling, still habitually practised then in some quarters. The aptly named Giles Silverside and his fraternity brother Samuel Garrett, both had vested interests in Kentish Town. If we consider Mr Silverside, and let it be said any preconceptions of his character as a caricature of the kind portrayed in satirical magazine Punch est. 1841 would be misplaced. In reality surviving evidence dictates otherwise, and he appears quite the contrary, an intelligent and determined businessman and a future master of the butchers company. He does of course share his surname with the silverside of beef; a cut of meat much favoured for salting that takes its name from the silver coloured sheath of fibrous tissue that separates each muscle.

Our first introduction to Giles in Kentish Town is by way of an insertion in Pigots 1823/24 Directory of St Pancras section, where he is listed as butcher in the company of five others all of whom traded in the area, no further information is given. He was in fact a meat wholesaler and

Figure 16: Garrett's Butchers, 62 High Street, Highgate, c1890.

retail butcher, and entered the butchers company as freeman in October 1792 where he was elected to Liveryman, then Assistant and finally in 1829 Master. We can be accurate in stating therefore he was already an accomplished and well-connected meat trader when business interests bought him to Kentish Town. His indenture of apprenticeship is held at the Berkshire record office in Reading, the county of his birth as he was born and grew up in Wantage before undertaking a seven year apprenticeship with butcher Robert Giles of Englefield. The exact location of his shop in Kentish Town has thus far not been identified, but is believed was in the area near to the Black Horse Tavern in Cain Place. A hostelry establishment of longstanding situated in the left fork section of Royal College Street that leads to Camden Road.

The shop was run by Richard and Stephen the elder sons of Giles Silverside as part of a wider business the latter named employing John Japp a journeyman butcher who lived in Fitzroy Place, Kentish Town. An early reference to the shop, together with family members involved, appears in proceedings of the Old Bailey in London. The case, one of simple larceny held on 14th May 1823, involved the theft of implements, namely a butcher's saw and sharpening steel valued at 4 shillings. This incident, as it would be described today, took place a fortnight previously when Thomas Perkins, Edward Bruce and James Grey all in their twenties stole the items from the shop, the property belonging to brothers Richard and Stephen Silverside. The culprits were apprehended by watchman James Halton in the early hours of the morning and taken to the watch-house in Camden Town having come from the direction of Kentish Town. All three men were found guilty and sentenced to six months confinement and publicly whipped.

The last reference to the shop appears to be in 1837, from an advertisement quoting Mr Silverside as a source of contact for letting a furnished apartment for a single gentleman in the best part of the village. The death of Stephen Silverside one year later raises the distinct possibility the shop closed soon after this event

Their father Giles lived and traded away from the area at 31 Paternoster Market, adjacent to Paternoster Row near St Paul's Cathedral, for centuries frequented by scribes, engravers and binders, buying books, prints and stationary. His other son Giles Silverside the younger, a meat salesman was responsible for the wholesale (carcass meat) market side of the business. In the eighteen-forties he lived above the premises at 6 Newgate Market, with wife Elizabeth and their son Charles, heir apparent to the family business. The Newgate Market, one of two principle meat markets then in London, its companion being Leadenhall.

It was from this market that Giles senior himself had cause to appear at the Old Bailey several years earlier in January 1817, when thirty year old William Jones was indicted for stealing fifty-six pounds weight of beef. He was apprehended after a struggle by Constable George Read in Creed Lane near Ludgate Hill. When questioned, Read replied 'A man had given me the meat'. He was found guilty as charged and sentenced to be confined for three months and publicly whipped one hundred yards from Newgate market. A wealthy man, accrued as head of a prosperous meat business and property ownership, Giles senior retired to Plaistow, Essex. When he died in February 1855, his will caused serious acrimony with certain family members and final settlement was delayed for two years. Soon after his son Giles opened a retail butchers shop at 172 Pentonville Road, Islington. The Newgate Market business was transferred to the new Smithfield Market in 1868, soon after the first two sections opened. By this time Giles the younger had also formed a partnership with his own son Charles until dissolved by mutual consent on 2nd January 1877. The family still trading there in 1884 as Silverside & Edwards, Meat Salesmen, 46 Avenue B, London Central Meat Market, as the market was formally titled.

Giles the patriarch was never resident in Kentish Town, unlike his contemporary Samuel Garrett, whose home address on admission to the guild in 1817 was 6 Prospect Place off Highgate Road. Presumably a Mrs Garrett living at Prospect House, Prospect Place, George Garrett publican of the Bull and Gate and William Garrett proprietor of a chandlers shop in Mansfield Place, all noted in the 1860s were related in some way. On the meat side the Garrett family had been around for some time at 18 Western Place, Pancras Road in the early eighteen hundreds, before Henry James Garrett undertook to open another shop at 62 High Street, Highgate. This shop under a later generation remained in the family until 1916, run by Henry Samuel Garrett with the assistance of his brother Herbert and their sister Florence.

That Giles Silverside and Samuel Garrett senior knew one another there can be no doubt, because both traded in Newgate meat market. In the year 1829 both were elevated to positions of prominence within the guild, Giles as already mentioned became Master of the Company in September and Samuel on the 13th of March to Liveryman. It would be inconceivable that one did not know the other, for both would have been obligated to attend official functions at the Butchers Hall then situated in Botolphs Lane where they almost certainly met. Like any other association the meeting rooms provided a haven to discuss and generate business, a habit frowned upon but nevertheless carried on, perhaps Giles and Samuel discussed their mutual affinity with Kentish Town.

Figure 17: Cattle grazing on Hampstead Heath.

There this particular story would have ended, had it not been for a man whose life involved observing and illustrating the world in which he lived. A talented amateur artist, Mr A. Crosby, believed to be a relative of bankrupt bookseller R. Crosby residing in Eden Place, painted and sketched scenes along the River Fleet, including by chance Prospect Place where Samuel Garrett resided. To assist his work, the artist wrote detailed notes and in one dated 23 December 1837 he mentions 'Mr Silverside, butcher, at present rents the grassland and barns from Mr Morgan a tenant farmer.' By good fortune Crosby's artistic pursuits unequivocally confirm Giles was embroiled in business affairs with at least one local farmer, and there may have been others.

It is conceivable the artist and the butcher may have been of slight acquaintance, but a more plausible explanation would be the artist knew of Giles by reputation from ale house gossip. How frequently Giles visited the area is impossible to say, wholesalers invariably used agents or drovers to act on their behalf; albeit in this instance we can presume he made occasional trips to discuss his mutual interest in the shop and financial affairs with sons Richard and Stephen and inspect the condition of his stock.

The Morgan farm Crosby mentions was sited at the junction of present day Caversham Road, Kentish Town Road, both land and farm owned by the College of Christ Church, Oxford. Succeeding generations of the family farming extensively throughout the area, including previously mentioned 'Shoulder Of Mutton' and 'Butchers' fields'. The latter being the smaller of two from the nine fields totalling sixty-eight acres rented by Richard Morgan in Kentish Town.

As the farming land gave way to bricks and mortar for housing in Kentish Town and elsewhere these rural activities disappeared, later generations of stock farmers including the descendents of William Hill, subsequently found themselves shopkeepers or proprietors of a wholesale butchery business, a natural transition given their experience and knowledge of animals. It maybe that a William Morgan, retail butcher, recorded in eighteen sixty at 70 Castle Road is such a case and is related to Morgan the farmers.

RHODES FARMERS

We can be positive in the descendent of the Rhodes farming family wherein the name Cecil Rhodes (1853-1902) is synonymous with colonist expansion in South Africa, his name perpetuated locally in an apartment block to commemorate his life survives in Goldington Street, Camden Town. His paternal ancestors William, Thomas and Samuel, are recorded as having substantial leasehold interests in farming and local politics. In 1733 William held the position of Overseer of the Poor, south division of St Pancras and in 1819 Thomas his son served as a member of the Select Vestry. The Rhodes family farm covered an area within the west side of Hampstead Road extending to Regents Park, with pockets of grassland and meadows elsewhere in St Pancras. One of the latter rented in 1804 by Thomas being Blue Barn meadow field in the vicinity of present day Queens Crescent market.

The extent of family farming operations judged from the sales catalogue, published on behalf of the executers of Samuels's estate, following his death in 1823. The sale in one Lot was conducted on 25th March 1823 by Winstanley & Sons, Paternoster Row and held on the Rhodes premises in the New Road near the Tottenham Court Road turnpike and directed at Cow-Keepers,

Butchers and Graziers of means. The list although lengthy provides an insight into the scope of farming, but more importantly conveys a sense of how one area of Camden earned its living at this period: "Valuable Leasehold Premises, comprising a large in closed yard, with extensive range of substantial brick-built and paved cow-houses, capable of containing 120 cows, stabling for 13 horses and suitable lofts, &c; a measuring house and dairy, granary, cart lodges, piggeries, and other conveniences, well supplied with water: with a commodious dwelling house adjoining the premises. Also the valuable lease of a grass farm, of 113 acres, with cottage, barns, cow lodges, stabling and outbuildings, situate and adjoining the high road leading from Pancras to Kentish Town, close to Regent's Canal. Live and Dead Farming Stock, comprising 99 excellent Yorkshire bred cows, many of which are adapted use: a fine bull, five valuable cart horses, breeding sows and pigs: several wagons and carts, three ricks of meadow hay, containing about 150 loads, a large quantity of manure, all dairy implements, a chaise and harness, a jaunting car and various other valuable effects, part of which are the farm".

By way of explanation a Rick is a stack of hay, straw or peas sometimes thatched for protection against the elements. The Chaise a corruption of the French chaire or seat, single horse or cob drawn vehicle widely used in the 18th and early 19th century, and Jaunting Car a two passenger horse or pony driven vehicle with back-to-back, sideways-on seating developed from the Irish Jaunty car. In 1890 Cecil Rhodes funded a monument in St Pancras Gardens to provide a permanent memorial to his ancestors.

MEATHIST MINIATURE

The old saying "Booth for Block and Bates for Pale" (meat & milk) relates to Thomas Booth and Thomas Bates both legendary for breeding shorthorn cattle. The renaissance of agriculture in the eighteen and nineteen centuries produce many such men who stimulated interest by exhibiting their success That for the majority of the population before the advent of energy giving sugar meant fat cattle to provide fuel for the body. One such animal the Great Durham ox, reputed to have weight 13/4 tons, the rolls of fat upon his back deep enough to hide a punch bowl.

3

SMALL INVESTMENT

The very early history of the retail food trade graphically demonstrates one of the fundamental differences between modern retailing and those of yesteryear. This was the strict demarcation line that once existed not only between different trades but within the trades themselves; that is to say for example although pork butchers were part of retail meat trading they only dealt at the beginning in one commodity. In much the same way poulterers and game dealers were not originally involved with the fishmonger's trade or the cheesemonger had not been absorbed into the grocery trade. The pork butcher, as a separate identity within the meat trade proper, emerges from the 13th century records of the guilds in London and York and reveals they were well established by this period.

From early on pig keeping assumed the lowly status of the working mans larder and the housewives kitchen sink, the latter an unkind reference, meaning in our modern society equivalent to a waste disposal unit. The cottager or cottage dwelling labourer for the most part left pigs unsupervised, their natural inquisitiveness and hunger encouraging them to scavenge. Their later widespread ownership, in every conceivable situation, and survival instincts exploited to the full down through the ages causing untold nuisance. In the well-known story 'Ivanhoe', Sir Walter Scott writes of Gurth the swineherd, son of Beowulph and his encounter with Wamba the court jester. From the banter between the two we learn something of the trials and tribulations of this occupation in medieval times. The patron of swine herders is Saint Anthony, from whence the name Anthony is given to the smallest and weakest piglet in the litter, and variously known by a host of other names such as a runt, dolly, etc.

For the prospective pork butcher this section of the trade offered practical advantages over his counterparts in the general sector; he could specialise on just one species which could be produced or purchased at relatively low cost. The enterprise was virtually self-sufficient, as pigs matured quickly and could be kept in confined spaces such as a hog yard behind the shop. Their diet was inexpensive and cost effective producing almost eighty per-cent of edible food; either processed as fresh or salted pork or bacon, meat products or a combination of all three items. In contrast sheep and cattle required grazing pastures, housing and a more elaborate slaughtering system, disposal of inedible waste material and different skills for processing hide, skins, wool and by-products. The down-side for pig-keepers was during warm or hot weather, pork deteriorated more rapidly than beef and mutton. In an attempt to overcome this problem, or in modern retail parlance extend the shelf life, the age old methods of pickling and dry salting were used. The joints and cuts packed in barrels or purpose built salting leads in cellars or out houses in the winter months, ready for sale during the summer months. The pickled meat was excessively salty at this period; nonetheless providing the correct procedures were followed would remain edible for up to two years.

The effectiveness of the curing processes (200 B.C.) was understood long before scientist were able to demonstrate the chemical changes that caused them, and even when this became possible they were far beyond the understanding of earlier generations of butchers. Men of practical minds, butchers knew little of osmosis, diffusion and sodium nitrate; instead half a pigs head was normally used as a starter organism; that is not to denigrate the early butchers, for man continues to put his faith in many things beyond his comprehension.

In common with all trades dealing in perishable foods susceptible to the vagaries of the English weather, an unusual mild spell could make the difference between profit and loss; butchers were hesitant to kill a pig because so much of the carcass required salting or worse became un-saleable. To minimise losses different cutting methods were used for winter and summer, combined with voluntary closing days. Nevertheless a regulation was imposed as late as 1752 in the reign of George III, making obligatory May 20th to August 24th a closed season (later rescinded) for pork; an historical precedent of regulations already set by the guild in previous centuries. So entrenched did this dictate become in the minds of successive generations of consumers, that long after refrigeration was universally available in shops and homes, mention of pork still evoked the once popular phrase, *'Never buy pork unless their is an 'R' in the month',* or as often described *'Dangerous Weather'*, a reference to the unhealthy deterioration of pork during unfavourable weather. The 'Now in Season' advertisement from Ernest Nicklinson that appeared in the 8th November 1890 edition of the *St Pancras Guardian*, was in part to promote public awareness that a new tradesman had opened for business in the area; although equally the arrival of new seasons pork, November to March being traditionally a time when pigs were slaughtered for market, as the colder months were essential for curing bacon until the 1850s when ice cellars began to be used for all year round curing. From 1877 mechanical refrigeration superseded ice for cooling and the emergence of the modern factory curing industry.

Outwardly successful in the initial years the business had spiralled into bankruptcy by 1893 forcing Mrs Jeanie Nicklinson to undertake the running of her husband's shop for the remainder of their tenure. Although in the intervening years her husband Ernest had begun trading from 88a Queens Crescent probably under the terms of a bankruptcy restriction order, nevertheless trustees were appointed and dividends paid.

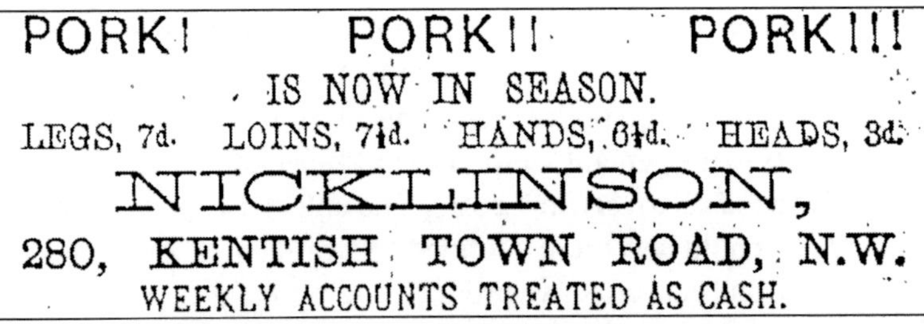

Figure 18: Nicklinson, 280 Kentish Town Road, 1890.

For the pork butcher and customer alike a vital part of his trade was the small scale production of pork products, cooked or uncooked, as distinct from pork sold fresh or salted. With a few exceptions, meat products were originally created from a necessity to utilize meat which otherwise was difficult to sell or keep wholesome, thereby providing a profitable alternative. These items were usually made on the premises and variously termed Small Goods, Made Up Goods or Small's; they were historically the prerogative of the pork butcher but subsequently available from other establishments such as general butchers, bakers, and grocers, ham beef and tongue shops, cooked meat and pie shops, delicatessen and a whole host of eating places. It is believed this aspect of the English pork trade was influenced by ships butchers and cooks who had accompanied their masters on travels to the continent as part of the entourage, and by French butchers well versed in the art of charcuterie domicile in London and other areas of the country from the 14th century.

It cannot be stated categorically, so we must err on the side of caution, but it seems likely amongst the earliest pork butchers shops to open for business in Kentish Town was that of Phillip and Edward Wilson. Others of course followed, however many engaged in the sale of other meats, a continual feature of this trade and others from the beginning of the eighteen-hundreds. This fact for the researcher makes positive trade identification extremely difficult; as does the emergence of the retail Porkman trade during the latter part of that century with the arrival of pork imports from America. The porkman's shop differed from traditional pork butchers in respect of being solely sellers of ready for sale pork cuts mainly from America packed in salt barrels, in addition to sales of hams, bacon, cheese and other associated items. Of no longevity in the evolution of the retail food trade, most were absorbed into the grocery and provisions sector and few illustrations exist.

J. RAYNER LTD

To compensate there survives a photograph from c1926 of Rayners window display showing some of the product range of the pork butcher. This particular photograph has featured in several local history publications not specific to retail butchers shops, and for this reason it is appropriate here to discuss the contents on display. The premises occupied a corner site fronting Queens Crescent and Basset Street, the latter a side road leading onto the market, the shop entrance flanked by two windows of roughly equal size. A refrigerated window base is unlikely, relying instead on wooden duckboards on a marble slab or other material with some form of primitive drainage for meat fluids. With regard to the actual display, the content and arrangement would suggest this is a show window mounted for a special occasion not a working window.

Figure 19: J. Rayner Ltd, 74 Queens Crescent, c1926.

The absence of a range of goods on the window bed, for example pickled meat etc, and the use of straw normally reserved for large displays of carcasses and poultry most frequently seen at Christmas, would seem to confirm this is the case. As intended, the 'Pork Much Cheaper' sign written across the window catches the eye, as do the seven pigs strewn with fresh parsley together with the ten sides of pork hanging. The liberal use of parsley as decoration in butchers' shops was quite common then and is still used today. As a general guide the higher the volume of meat on show, particularly pork and pork products, the colder the month; with set piece displays for photographic purposes these were very often removed immediately afterwards. At this late period in time this firm dealt in meat other than pork, so beef, lamb and poultry were confined to the opposite window fronting the market side and sale of frozen meat from its neighbouring branch at number 58 in this road. Nonetheless the photograph captures for posterity a brief insight of the butchers' craft of small goods production. The majority of products on display with prices have been identified in the accompanying chart.

The selection of cooked meats and fresh sausages on show will certainly be familiar to consumers of today, the retail prices unfortunately will not. The Heinz 57 Variety tomato sauce on the float an American product first registered there in eighteen sixty nine and available in this country seventeen years later. A curiosity today, the chicken and ham roll contained a variety of cured boneless meats: chicken tongues, veal, pork shoulder and beef. The recipe then all perfectly legal and remaining so until recent times.

There was great emphasis on informing potential customers the pork on sale came from English dairy feed pigs, a diet high in whey and other milk by-products; an unjustified inference that the imported article blamed for the demise of pork butchers was feed on swill and vegetable matter and therefore inferior to home produced. The quality and quantity of Dutch imports of pork and bacon at this period were becoming increasingly worrying to the trade. The sale of additional products derived from beef, quaintly titled in Victorian times as 'fresh edibles of the bovine', as with mutton and poultry in pork butchers had for many years past been the norm in London and throughout Southern England very few still specialised exclusively in pork by this date. The products on display in Rayners shops were either made at their sausage factory located behind their Hampstead Road branch or supplied by outside contractors, as large scale production of small goods using machinery was then well established.

Description - pre-decimal prices

per lb/each			*per lb/each*		
Cooked Ham	2s	8d	Steak & Pie		3d
Chix & Ham Roll	1s	4d	Meat patties		2d
Brawn		10d	Pork Pies		3d
Ox Tongue	3s	6d	Pork Luncheon Meat	1s	4d
Pressed Beef	1s	2d	Beef Luncheon Meat		10d
Veal Galantine		u/k	Cambridge Sausages	1s	2d
Black Pudding		8d	Beef Sausages		u/k
Smoked Ham	3s	4d	Pressed Beef	1s	2d
Cooked Ham	2s	8d	Pork Loin	1s	6d
Leg of Pork	1s	3d			

As to the family genealogy and detailed business history the nature of records as have survived prohibit a true and concise chronological picture unfolding and for this reason the following is a miscellany of facts that have been judged to contribute to the story behind the photograph.

The name Rayner first comes to notice in north west St Pancras with butcher Thomas Rayner at number 8 Murray Street, Camden Square c1860, his trade confirmed in the vestry minutes by the annual renewal of a slaughterhouse license granted in his name at this address and corroborated in St Pancras rate books that reveal butcher Thomas James Rayner was occupying the premises and still trading here in 1910. At an earlier period Thomas dissolved a business partnership with Charles Cooke, trading as oil and colour men under the style of Rayner & Cooke at 6 Murray Street. The latter an isolated statistic but indicating an inclination for business partnerships that in hindsight revealed a foretaste of future business strategy and may give credence the following family are one and the same.

The first appearance of Rayner in Queens Crescent commences in the early eighteen seventies, via number seventy-four through a partnership arrangement with butcher Frederick King. The beginning of a bewildering number of business partnerships and titles trading from five shops in this market road and at other locations involving Messrs Rayner, Rayner & Co, Tower, King and Hardwick. The last named, Frederick Hardwick, born in Kings Cross and known to be in his middle thirties at the time of his business associations with the family who were also active on Smithfield Market as wholesalers in partnerships with other market traders. Retail shops in other areas bearing the Rayner name include Euston, Camden Town, Holloway, Islington, and Paddington. It was not until 1912 the Rayner business became incorporated as a sole identity; preceding this event George Rayner instigated numerous partnership dissolutions, typical of which in 1907 Rayner and Barnett carrying on business as Rayner & Co, 49 Goodge Street, Tottenham Court Road. Regretfully other than the company name Rayner Ltd and registration number the official records have been destroyed. On the 1st of April 1966 Rayner & Co (Pork Butchers) Ltd held an Extraordinary General Meeting at Verulam Buildings, Grey's Inn, London, where the following resolution was duly passed: "That the Company be wound up voluntarily and a Liquidator be appointed for the purpose of such winding-up".

There is no evidence thus far that any member of the Rayner family lived in Queens Crescent, although it is known the family held the lease on property numbered seventy-four in the road and possibly other premises; as was common the rooms above the shops providing accommodation for employees. Turning to other sources reveals a member of the Rayner family engaged in the wider issues affecting the trade. The National Federation of the Meat Traders Association, Pork and Bacon section formed in 1919 from The London Butchers & Pork Butchers Society founded in 1820, had within its London membership James E. Rayner, a staunch advocate of the trade. In 1927 he was co-opted to represent the view of London traders on a sub-committee appointed by the Federation at the request of the Ministry of Agriculture, to investigate the possibility of producing a breed of pig commercially suitable for their trade and demands of the consumer, in short the butchers' ideal pig. From a professional point of view, a task he was eminently qualified to do judging from the pork carcasses purchased and on display in the photograph. Although to the experienced eye slight trimming of back-fat from the sides of pork can be detected but nevertheless the leanness would find favour even in today's health conscious society. This weight and standard of pig was primarily used for joints, chops and steaks and referred to as porkers or Londoners, the latter term meaning simply acceptable for the London trade who preferred lean pork normally from a Middle White breed, a heavier weight of pig was used for production of pork and bacon products.

In recognition of his service and loyalty on behalf of the pork butchers trade James Rayner was voted chairman elect of the pork and bacon section in 1933. Unfortunately under distressing circumstances for himself his tenure is remembered as unique in the history of the section. He would normally have been installed as Chairman at the 1934 bi-annual conference taking place that year at Plymouth. However James became seriously ill a short time before and was invested with the chain of office, purchased in 1924, at his home, the presentation taking place in the presence of family and officials at 7 Wyndham Crescent, Dartmouth Park Hill. The condition of his health was such he was prevented from attending any meetings or functions until his period of office was due to expire. It is unknown if this illness affected his capacity to conduct business affairs, but within a short space of time the company had ceased trading. With the departure of Rayners from Queens Crescent in 1934, one of the last practitioners of the pork butchers craft in Kentish Town had gone. A fact of local interest, it did however mirror events elsewhere most notably in London and the south of the country; although this is not to imply they were a spent force especially in the industrialised areas of the Midlands and North of England.

The north, with a higher proportion of labour intensive industries like coal mining, ship building and steel works and especially employment of women in the textile trade had a much stronger customer dependency on the services of pork butchers, and for this reason the trade persisted until recent times, albeit currently only a handful remain. There were a number of contributory causes for the demise of this retail side of the trade, foreign imports of salt pork, bacon and hams and depletion of the national swine herds from epidemics, and not least the Englishmen's deep seated preference for beef and mutton, but not pork. Substantial imports of these other meats arrived in the frozen state from the 1890s; widely available and reasonably priced it meant families meals were less reliant on pork meat which unquestionably lacked consumer confidence, and home produced beef which was relatively expensive. Paradoxically it was the introduction of mechanical refrigeration, intended to ease the burden of financial loss, which inflicted most damage. An innovation, hailed by many prominent trade personalities of the day as 'the saviour of the pork butchers trade and oblivious to the fact it was a two edged sword which had a devastating impact.

The trickle of competition from all quarters became a flood, and traditional butchers had no compunction in adding regular supplies of pork to their meat range. Faced with the inevitable, a sizeable proportion of pork butchers either became general butchers or ceased trading. Worse was yet to come, for in the months prior to the Second World War the department of the Board of Trade responsible for planning the food defence strategy announced what amounted to the coup de gras. If war came the bacon market would receive first priority on the available supply of pigs, few if any would be available for pork butchers. The government attitude was at first uncompromising, change to general butchery or bacon curers or shut up shop for the duration that despite exemptions signalled further losses on return to peace time conditions.

The rate of closures for any given period is difficult to assess because no accurate government or local authority statistics are available, other sources of information such as pork butchers associations had mixed membership and not all pork butchers joined trade organisations. Certainly from the eighteen nineties we see a downturn in directory entries covering the whole of London and in the locality of Kentish Town. From approximately twenty six pork butchers trading within the district at this period only five had survived as such beyond the turn of the century: Rayner,

Queens Crescent, H. Ritchie, and P. N. Brazil Kentish Town Road together with Charles Roach, Fortess Road and John Ulm, Ferdinand Street. In 1903 only one hundred and sixty pork butchers are listed for the whole of London, a figure treated with caution as pork butchers continued to be listed as such although in reality the vast majority were trading as general butchers.

SELL THE SIZZLE

Of all the meat based products, the sausage although available from other outlets, is most closely identified with the retail butcher. It began its commercial popularity as the stock in trade of the pork butcher, as the pork sausage was and still remains the nation's favourite, although it is said Scotland still favours the beef sausage.

The enduring popularity of this product with butchers can in part be attributed to its simplicity in production, entirely hand made (now rare) or minimal machinery required; for the consumer the culinary convenience as a snack or part of a main meal and the various cooking options available. A representative example of the types commonly made is fresh, cooked or dried. As to the sausage itself, the basic concept is simply a filling normally comminute meat enclosed in a casing either natural or synthetic.

The origins of the sausage are somewhat more problematical because the ancestry of good ideas; its discovery or invention and birth place has many claimants and thereby lays the difficulty. Did Neolithic man discover the concept by accident or a deliberated act, or are its origins in the ancient civilisations of Assyria and Samaria or China, Italy, Greece or Germany, as have been claimed. It has been reputed sausages were made and eaten in the ancient communities of the Assyrians living in the city of Tyre on the coast of Lebanon and Sumerians in Babylon 5,000 years ago. Greek history provides one of the earliest authentic references, in 850 BC Homer's epic adventure 'Odyssey', when he alludes to a man standing beside a great fire cooking a stomach filled with fat and blood.

Later the sausage raised to literary prominence in '*The Orya*' a play written in 500 BC. There are countless theories and myths surrounding the discovery or invention of the sausage, alternating between farce and the ridiculous. The most likely and persistent assumption that must be given credence is our Neolithic ancestors were responsible, and even here there are differences of opinion; that Neolithic man perhaps concluded from experience meat stored in animal intestines kept longer, or meat so stored accidentally became dried and smoked by heat from the fire thus inhibiting deterioration. At the risk of offending the reader another theory is that quite by chance partly digested food in the intestinal tract of animals hunted for food was discovered on cooking not only to keep longer but could be stored in its own natural container. Its true origins may never be traced, but wherever the sausage originated the Romans are justifiably credited with spreading its popularity to the four corners of their Empire. In Britain, in the absence of evidence to the contrary, our present understanding is the Romans were the first to introduce not only the sausage in all its forms but also the name derived from the Latin salsisium meaning something salted; which gives the reader a clue to the type of sausage they originally made heavily salted and dried, two of the oldest forms of meat preservation and an essential consideration in relatively hot countries. Other ingredients included spices and herbs to mask the odour and improve the flavour of mainly coarse and unwholesome meat, with pine kernels, and berries to act as a binder and absorb fat.

From the very beginning the intestines of animals were used in the making sausages and easily digestible, they complement the ingredients perfectly. The sausage skin, or proper trade title casing, are derived from cattle, sheep and pigs, each given its own trade or technical terminology for various kinds of sausage: Bungs, Runners, Weasands, Fat Ends Hogg or Ox casings and many more. Some of the finest sheep casings in the world came from Russia the renowned Kalmucks, and from Afghanistan and Mongolia, their quality attributed to the climate and other factors. The small intestine of the pig varies from 50 to 65 feet and the large intestine is 12 to 15 feet, which goes some way to explain how butchers were able to produce the beehive looking spirals that so fascinated Victorian photographers. In the 17th century sausages were first twisted into 'Lincks' with the exception of the Cumberland sausage which by tradition until recent times was not linked, being sold by length, hence the continual spiral from which portions were cut. The linking of sausages enabled them to be sold at a specific number to the pound weight, eight pork, six beef etc and approximately twenty chipolatas. This was also a convenient way for butchers to distinguish between unmarked varieties similar in colour. Customers are often confused with regard to chipolatas and unless advised otherwise will be sold a miniature version of the standard pork variety. The chipolata is not just a thin sausage, but should be made to a proven quality recipe from high grade meat, seasonings and superior lamb casings. Linked sausages provided more than just ease of serving customers, butchers with a more manageable product were able to carry out further processing by hanging small amounts of sausages in chimneys to obtain a smoked flavour.

The term 'home-made' often seen on posters and shop windows refers to the centuries old tradition when butchers lived on the premises and kept pigs in the back yard, which were fed with a mixture of shop and domestic scraps supplied by customers. The connotation being that anything made at home is less likely to be adulterated and therefore of superior quality, an assumption not always fully justified. A more appropriate term today is 'own make' or 'brand'.

ELIZABETH ATTKINS

We have pictorial evidence in Elizabeth Attkins' pork butchers shop, at 55 Highgate High Street (originally No 9) with sausages hanging from the rail in the widow display, of whom it was said had a reputation for the very best quality pork sausages in the district. The Attkins family of retail butchers settled in the Highgate area before the turn of the 19th century. Commencing with the Samuel Attkins family and extending to several branches of succeeding generations in trade as butchers and poulterers from locations in this road until the 1970s. Samuel is noted here in 1805, and later in a reference to his shop in 1822 during a

Figure 20: E. Attkins, 55 High Street, Highgate, 1885.

burglary trial at the Old Bailey. A witness mentioned the shop "We got two pounds of beefsteaks from Mr Attkins the butcher and then went to the Rose and Crown to have them cooked with eggs".

A local worthy Samuel was involved with other individuals in 1827 as Overseers of the Poor in the matter of a £1,800 bond and contract to build an additional wing at Hornsey workhouse. His concern for the less fortunate sorely tested several months later having to attend the Old Bailey court in person together with his daughter Anne and James his nephew. The defendant Anthony Bernard, a French national, indicted for stealing amongst other items five sovereigns and five £5 pound bank notes. One of the stolen bank notes unwittingly used by a customer to purchase meat in Attkins shop and handled by three member of the family. When and where Samuel Attkins was born has not been discovered, but he died at his Highgate home in February 1838. The Elizabeth Attkins' shop was home and work place for butchers boy Edward Attkins aged 18 years, son of nearby relatives and Robert Dawkins a live-in pork butcher and Elizabeth Susan her 18 year old daughter of unknown employment.

Of greatest impact, George Attkins benefiting from an inherent aptitude nurtured in a meat trade environment and during his tenure at 55 in the High Street, assisted by his wife Katie butchers cashier and their sons Stanley and William, assistant pork butchers. An all-round butcher, but specialising in pork, George reared his own pigs and competed in competition at the Royal Smithfield Christmas shows held in the Royal Agricultural Hall, Islington.

A talented pig breeder housing dozens of pigs in sties situated behind Townsend Yard near his shop, he attended numerous shows between the years 1892-1907 and won many cash prizes, and was placed in every class he entered during that period. He specialised in the Tamworth, Black, White and Cross breed pigs, the quality of which was undoubtedly responsible for his high reputation as a true Master pork butcher among his contemporises and customers in Highgate and surrounding districts.

From early on it became apparent independent slaughtering facilities for each Attkins enterprise was not an option, and a solution was found by using a communal slaughterhouse situate at the rear of York Place, a run of houses previously known as Watch House Row accessed via South Grove on the south side. From 1858 the vestry minutes record an officially licensed slaughterhouse behind York Place and Mrs Elizabeth Attkins shop and named as the licensee on numerous occasions. The slaughterhouse license granted annually subject to inspection (more frequent if required) on one occasion the sanitary committee gave a rating of satisfactory. In a coincidence of events, in 1893 the name York Place was officially abolished and the slaughterhouse issued with a new registration number 223 and relocated to 61 Highgate High Street. The licensee for this address in 1901 still Mrs Elizabeth Attkins, although slaughtering restrictions now stipulated pigs only.

Figure 21: Palethorpe Ltd, Highgate 2008.

The recipe of their celebrated pork sausage it was said still unchanged, for in a country whose knowledge of the world if not its empire was founded on the search for rare and exotic flavourings, in the days of dietary conservatism with a positive aversion to anything in meals remotely foreign sounding or tasting was baffling. Nonetheless, most strongly disliked by the food shopping public was highly spiced flavouring in sausages and meat products. To this end in the nineteen thirties advice for butchers wishing to produce their own sausages was still unequivocal and precise, "a customer who needs his food highly seasoned has the cruet at his elbow, whereas an over-seasoned article will be quite useless to the majority of customers". A general attitude accentuated by two world wars, and the bland utility or toothpaste sausage still prevalent in gloomy post second world war Britain, only gradually changing from the nineteen sixties. Although understandably the once popular German sausage, a commercial casualty of propaganda was hastily renamed Empire sausage on the outset of the Great War; this superior large luncheon type also known as Windsor Roll; the Colonial black pudding lost its prefix in the nineteen thirties. The saveloy because of their cheapness are an exception, the name is frequently explained probably correctly as an English corruption of cervelas (cervello-brain) because the saveloy could contain any part of the pig except the brain. Of the firms that became high profile sausage manufacturers in their day: Brazil, Bowyers, Drings, Lyons, Palethorpe, Richmond, Harris of Calne, and T. Wall are a few of the names.

Of these a consideration for historical heritage and architectural integrity in Highgate High Street has ensured a sign advertising Palethorpes Cambridge sausages has been preserved. The sausage advertisement in white lettering on a royal blue background clearly visible high up on the east facing wall. The Midlands based firm was founded by Henry Palethorpe born in 1829 in Tipton, Staffordshire West Midlands, where a road in the town bears his the name. By the 1920s trading as Palethorpe Ltd, sausage manufacturer, headed by his son Charles Henry. The family lived in some grandeur in their own castle aptly titled 'The Sausage Castle' by the local population. An extremely wealthy man Charles died aged 66 years in November 1922, leaving a personal fortune of £231,565, from which a considerable amount was given in bequests and trusts to the City of Birmingham institutions. However the most widely used and enduring legacy if true, is to the English language. As the company lay claim to the term 'Snorkers' a colloquialism it is said first used during the second world war by the Royal Navy submarine service for Palethorpe tinned sausages. This be as it may, the business rivalled many of the major players in the production and sale of quality sausage and other meat products. The volume counted in tons requiring the ownership of special railway vans using the network of G.W. & M.L. railway for distribution. The Palethorpe vans painted on both sides with the company name and a pack of sausages were in service until the nineteen-sixties. In 1990 this long established firm was acquired by Northern Foods, a conglomerate of food companies ranging from confectionary to condensed milk. A cost saving rationalisation programme by the parent company in 2006 to concentrate on the most profitable parts of its business resulted with Palethorpes being advertised on the financial market for sale.

The most famous of all however, and still a household name, is T. Wall & Sons Ltd, the founder Richard Wall a Londoner started as butchers apprentice in c1786-90 to master pork butcher Edmund Cotterill proprietor of a small shop in St James Market, London. He eventually became a partner taking sole charge of the business after the death of Mr Cotterill. In 1812 the firm received its first Royal Warrant as purveyor of pork to H.R.H. George Prince of Wales, later the Prince Regent and eleven more Royal Appointments were added by 1961. The Walls reputation for pork sausages and other products spread throughout London, and to meet demand additional premises were acquired most notable of these was 130 Jermyn Street; successive generations of the family continued to expand the business opening additional factories. A bid was made for the company by Mac Fisheries, the fresh fish shop chain owned by William Lever (Viscount Leverhulme), the purchase price of £120,000 however overstretched Mr Lever's finances and in 1922 both Walls and Mac Fisheries were transferred to his parent company Lever Brothers Ltd, and eventually in 1929 to Unilever. By the early 1930's the Walls company was reputed to have a fleet of 45 vans and 40 trade tri-cycles supplying meat and sausages to over two thousand shops, about this time they introduced bacon to their range and would later be the first to offer rindless bacon. On the first day of January 1986 Walls joined another famous meat company and became Mattessons Walls as part of the Unilever Empire.

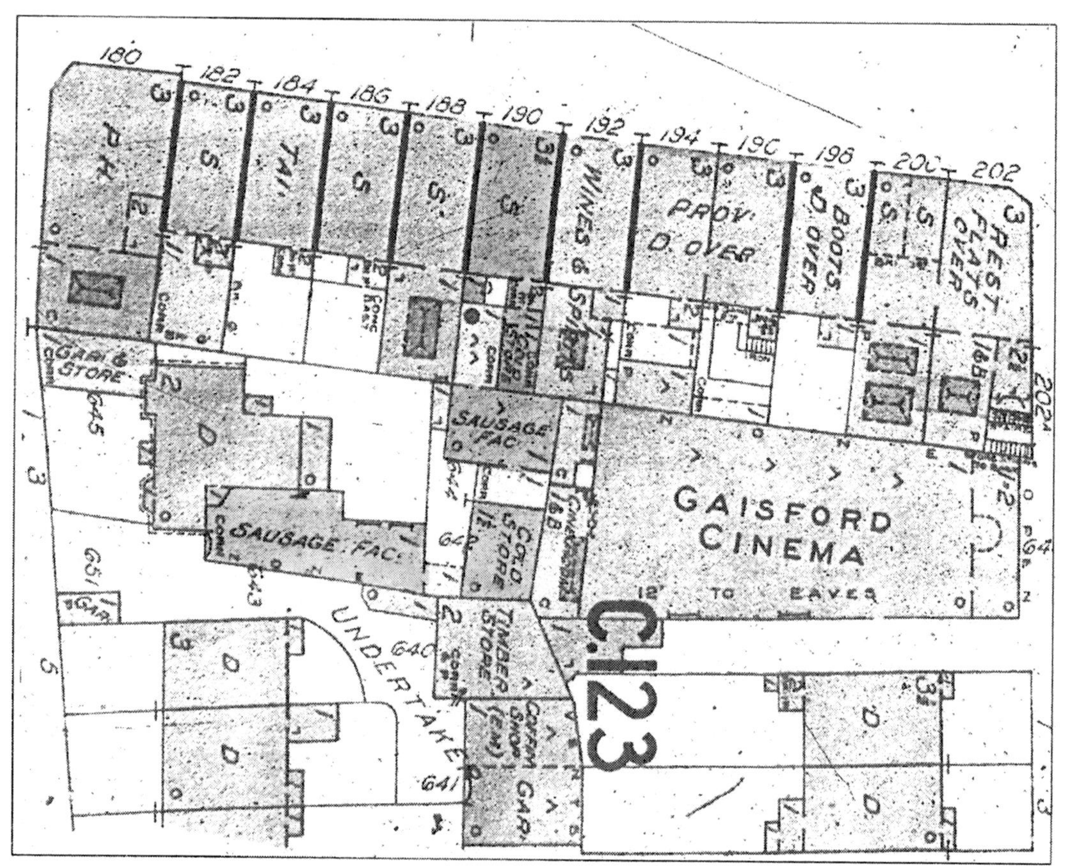

Figure 22: H. Ritchie, Shop and Sausage Factory, Kentish Town Road, Goads Insurance Map, 1954.

At local level in the main shopping thoroughfare of Kentish Town there existed for many years a considerable sausage making enterprise operated by high street butcher Harry Ritchie. In Goads insurance map can be seen his shop and the factories together with cold stores. The sausage production used to supply the Ritchie shops and other retailers and caterers.

CORRIGAN BROTHERS

Figure 23: Paul Corrigan c1985

Kentish Town would have to await the arrival of the Corrigan Brothers, who as far as is known are the only sausage and meat product manufacturers with a national reputation outside the area. The business was founded by brothers Paul and Frank from Newbridge, County Kildare. Frank arrived here in 1953 and first worked for Cramer Butchers York Way, N7. He later became responsible for running the Cramers branch shop in Brecknock Road. Meanwhile Paul had returned from a job seeking mission abroad and taken over his brother's duties at the York Way shop. The two brothers later opened on their own account at 63 Camden Road attracting customers from the large Irish community within the district. They quickly gained a reputation for making meat products that ranged from Irish white pudding to salt beef and home cured Limerick ham; the jewel in the crown so to speak was their pork sausages, attracting customers from far and wide. The high standard and quality confirmed when awarded a bronze medal in a national competition held at the Agricultural Hall, Westminster.

The judges were husband and wife team Fanny and Johnny Craddock, celebrity food and cookery personalities, hosting their own television program. The pork sausage recipe, reputedly worth thousands of pounds to commercial competitors, locked safely away in the local bank. By 1974 the number of branches had risen to ten, Royal College Street, Dartmouth Park Hill and Chetwynd Road, amongst them. In 1977 the business was incorporated as a private limited company with registered office at 18 Brecknock Road, the butchers shop previously owned by Philip Cramer.

Thereafter the company progressed from predominantly a family butchers into a highly respected meat products company; whilst still retaining the Irish customer base and trading policy of sourcing only the best quality English and Scotch beef and pork from Ipswich and Norfolk poultry. At the centre of Corrigan's successful business was their effective commercial exploitation of their prize winning home-made pork sausages and meat products, enabling the company to pre-empt the decline in retail butchers shops by transferring the bulk of their operations to manufacturing. By 1990 most of the branches had ceased trading and were leased or sold to existing employees; the Camden Road branch passed to Mr D. Doyle. The company retaining only a few of the original shops, using the following trade titles: Corrigan Bros (Butchers) Ltd, and Corrigan Butchers (Holloway) Ltd. In 2003 Paul Corrigan died and soon after Frank retired to his native Ireland to live in Sligo. At a General meeting of the members of Corrigan Bros (Butchers) Ltd, held on 6th November 2008, a Special Resolution was passed that the company be wound up voluntarily and a Liquidator be appointed at which time family member Mr S. B. Corrigan was a Director.

SWISS COTTAGE SALAMI

Of great significance to the development of pork meat products in this country from the second half of the nineteenth century has been the influence of German butchers, who are acknowledged experts in this field; and none more so than Michael Katz who, together with his future wife Ilse, arrived in England during August 1939 to escape persecution and probable death from the Nazi regime, a fate which sadly befell relatives left behind. The son of a master butcher, Michael was born on 21st October 1918 the owner of the largest chain of butchers' shops in Cologne. After leaving school he began an apprenticeship, working in the family shops and factory under the supervision of his father. His natural abilities and determination were rewarded on passing his trade examination with the highest marks.

The oppressive Nazi regime however meant that life for Michael, and the family of Jewish descent was becoming untenable and the business was forced to shut down through lack of customers. The result of a deliberate policy to force the closure of Jewish businesses; without any hope of a long term future Michael and Ilse made their way to England. So all embracing was the National Socialist Party quest for a totalitarian state there was an attempt to give the many German meat and sausages products standardised names throughout the Reich. The butchers of The Grand Advisory Council of Reich Guilds, at a meeting in Cologne in April 1936, dismissed the idea as totally unworkable.

Soon after his arrival in England, Michael found employment with P.C. Ford butchers in Bristol, a piece of good fortune brought to an abrupt end when war time regulations concerning foreign nationals decreed he had to leave the city. A major setback that in turn produced the opportunity for Michael to use his undoubted expertise in the manufacture of meat products, when employed as second machine man making sausages for R. Gunners Ltd, butchers at their factory in Lever Street, EC1. After a spell here a more financially rewarding and challenging position became vacant as production manager for butcher F. Kaye & Sons in Mitcham Road, Tooting.

In 1941, Ilse managed to find employment at Silver Stores, Northways Court, Swiss Cottage and accommodation, a one room flat in Homefield Court, Belsize Grove NW3, the couple's first real home, but life was still far from easy. With a shortage of money and household items in short supply, Michael put his sausage making skill to good use as he relates here: "I made salami in the kitchen from beef fat and goat meat flavoured with plenty of garlic. I minced the meat while Ilse stuffed the skins wearing a gas mask, because being pregnant at the time, she could not stand the smell of garlic. I would sleep by the cooker until the alarm went at 1 am, indicating that they were ready. I would then go to bed. Ilse would wheel them, in a pram, to her shop for sale the next day. People looking into the pram to see the baby had quite a shock! The first week I made salami I put four pounds in the bank and seven pounds the second week, this was in 1943/4." In this instance, the delicatessen shop could truly be described as having real 'home made' salamis

Figure 24: Scot of Bletchley, Logo, 1968

for sale. Obviously, no serious attempt to run a business was intended, but it did provide extra funds and little did neighbours realise the industrious Mr & Mrs Katz in their midst would one day be co-owners of Scott Meat Products of Bletchley, established in 1961 and among the most successful meat product companies of post war years, with a turnover of 14 million and several thousand employees in their first decade. In no small measure due to the founder Michael Katz for whom worker, employee relationship included a happy environment. As he recalled during his frequent visits to the processing plant 'When they are pleased with me they sing happy songs, when they are angry with me, they sing me a rude song'.

POOR MAN'S RESTAURANT

Such was the diversity of edible products made from the pig it was said pork butchers had an advantage because everything could be utilised except the curl in a pig's tail or the squeal. Although exceeding the bounds of advertising decency, an enterprising American pork meat company went further by issuing a macabre gramophone recording of a pig's squeal, boasting we even use the squeal; equally insensitive was the term walking larder used throughout the trade. Unbelievably the proverb 'You cannot make a silk purse out of a sow's ear' was disproved when another American company spun a small piece of silk after first dissolving bristles from a pig's ear in caustic soda.

For pork butchers the production of these items provided a useful outlet for the cheaper cuts, meat trimmings and offal which otherwise might remain unsold, indeed such was the popularity for many butchers this trade accounted for the major part of their business. Always more popular in working class districts, especially in industrial towns class snobbery ever present, pork butchers became known as the 'poor man's restaurant' or 'pig and pie shops'; best selling lines all sold hot and accompanied with pease pudding or mash potatoes were sausages plain pork or saveloy, pork pies and faggots, the latter re-named savoury duck in some quarters to improve its image and disguise its true identity and ingredients. The new name and dubious ingredients or not, it tickled the taste buds of one culinary commentator who described it as a 'very toothsome article'.

In the comforting veil of nostalgia it is often forgotten by older generations why these items were so popular when there was plenty of prime meat for sale, the answer in the majority of cases was low income. The pie man or picnic pie man was a familiar sight in the streets of London, selling a wide variety of pies made with beef, mutton, pork, eels and all kinds of fruits when in season. The dubious nature of the meat ingredients included by some vendors lead to shouts of 'Mee-ow' or 'Bow-wow-wow' whilst touting for customers. An itinerant way of life that began to fade from the London scene during late 1850s, as pie shops and other eateries increased in number.

Small goods are the forerunner of today's take-away meals, many both in name and nature may seem strange and unappetising to modern day consumers. Yet at meal times in days gone by they were an essential part of the daily diet. There was considerable regional variation in customer preference and colloquial names, and if the food was sold hot or cold. Any list of these goods would be ad infinitum, as one contributor mused "the pork butcher should remember variety is charming" in a practical treatise on the subject. The following examples are just some of them; pluck pie, pigs fry, corned leg of pork, chitterlings, faggots, bath chaps, collard head, sweet breeds, mother maws, pork cheese, pigs snout, and pork pies, and saveloys. In explanation the term 'mother maws' in this instance colloquial butchers slang for pig stomach, the word 'maw' meaning the mouth, jaws, throat, or stomach of an animal, especially a carnivorous animal that devours food greedily.

Items like pigs ears and tails, trotters, cooked rinds, bacon and ham bones and pork scratchings were the equivalent of today's snack foods comparable to burgers, hot dogs and such like. The scratchings still available today alongside its direct descendant the potato crisp. From the many brand names Walker's crisp were established in the 1960s under the direction of Gerry Gerrard, a member of the Walker family; a retail butchery business with twelve outlets based in the Leicestershire. The Smiths crisp brands were first produced in 1919 by Frank Smith, the business incorporated in 1929.

The majority of products then were not snacks or Sunday afternoon outing treats, but providing a midday or night time meal: hot-eels, pickled whelks, oysters, sheep's trotters, pea soup, fried fish, hot green peas, kidney pudding, boiled meat puddings, beef, mutton, kidney and eel pies, pasties and baked potatoes.

As local Kentish Town resident Ena Baker recalled from her childhood days in the early nineteen hundreds 'On Mondays mother did the washing in a big copper with a fire under it and she gave me shilling to get lunch at the local mashed potato and sausage shop'.

Correspondent Tony Obrist, living in Liverpool Street, Kings Cross, during his childhood days writes: 'Towards the east end of Cromer Street in a row of shops on the south side was Matthes which I believe were butchers of German extraction. It was the custom in the shop to be provided with a dish and fork and you could select your own cuts of meat out of the window. When buying chops, my mother always asked the butcher to saw, not chop the bone. My dominating recollection is of a barrel of steaming pease-pudding standing by the counter a tasty and very filling food, of which a penny worth was more than a small boy could manage at a sitting'.

A multitude of eating houses provided varying standards of food and drink with prices to suit, Kentish Town as elsewhere had its share of dining rooms and coffee-houses. In contrast to our present society the oven was non-existent in poorer homes which lacked even the most basic of domestic cooking facilities; so eating out was less of a social pleasure and more of a necessity, hence the growth of any establishment or vendor be it butcher, baker, tavern, eating house or hawker that provided ready prepared food hot or cold. A state of affairs that existed until fairly modern times, when many families shared cooking facilities or in the case of single room occupants had none.

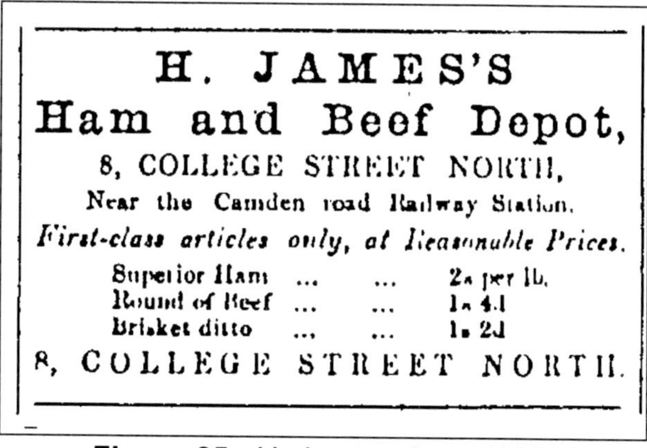

Figure 25: H. James, June 1866.

Another contributor to the 1983 *Camden History Society* essay competition, Mrs Asbolt writes: 'In the corner of Regina Street, the butcher was Mr Knapman and at Christmas some families shared a turkey and the butcher used to cut them in half. At the top of Queen Street was the bakers, on Sunday people came from all over to cook their dinner in his oven you would see the meat resting on the top of the potatoes, and it only cost a couple of coppers'. A service that butchers found of advantage in operation of their own business through the centuries; the local baker's ovens a convenient place to cook their pies. The proprietor of the butchers shop referred to by Mrs Asbolt was Edward George Knapman at 15 Plender Street, Camden Town.

From an earlier period, on Sunday 27th November 1889, between one and two o'clock in the afternoon to be precise, we have another account from the note book of an investigator in the services of Charles Booth the Victorian reformer. The investigator known only by his initials E. A. gives the following account, whilst recording social conditions in Kentish Town. 'A woman of perhaps thirty or so hurries home with a dinner from the baker's. At least one family in Litcham Street was going to dine royally off a substantial joint of mutton and baked potatoes. The women looked proud, and I am afraid, the neighbours envious'. This particular street gained notoriety for dreadful overcrowding and filthy living conditions until redeveloped by the St Pancras Housing Society. The road was renamed Athlone Street after Princess Alice, Countess of Athlone who in 1933 opened the first block of flats erected. In a perverse twist of irony, among the first tenants was the landlord, partly responsible for creating the slum conditions in this street and in other areas of London.

Had the un-named women in Litcham Street tried to avail herself of this particular cooking service in earlier times the following restrictions would have applied, because under the "Baker Act" then in force, stated it was unlawful for bakers to deliver baked meats, pies etc, after half-past 1 o'clock on a Sunday which have been cooked for customers. As Mr Mead, the proprietor of a baker's shop in St Andrews Street, Seven Dials area of Kings Cross, found to his regret in 1867, after being fined 10 shillings. The shop of the unfortunate tradesman had been under surveillance by Messrs Tinkler and Diamond members of the Amalgamated Journeymen Bakers' Association, who reported searching a dozen or more customers supplied with baked dishes, among them roast loin and leg of pork.

Enduring and most frequented of all eateries and currently enjoying a renascence, are the pie and mash shops, the meat pies often supplied by local butchers. In medieval times meat pies were called 'coffins', echo's of the infamous Sweeny Todd the demon barber of whom it was said made pies from the grizzly remains of his victims. Another unusual name origin was kit-cats alias mutton pies made by Christopher Cat, a London pastry cook. From which the Kit-Cat club founded by Jacob Tonson in 1700 derived its name. The club was frequented by the elite of writers and politicians of the day until forced to close down in the nineteen twenties. There was also the wonderful Dickensian sounding Farthing Pye House famed for its mutton pies and mutton dumplings, sold for one farthing each, sited in opposite Portland Road Station; the farthing was withdrawn from circulation in nineteen sixty its monetary value worthless. The indomitable Castle pie and mash shop establish at 229 Royal College Street, in 1934 still provides the customer with this most traditional of English meals; trade descendants of another Eel Pye House at 12a Kentish Town Road run by Elsie Antik in 1893.

The eateries were not entirely the domain of pigs feet and pease pudding establishments for in 1894 J. Lyons & Co, opened their first tea shop in Piccadilly, one of a hundred outlets across London that would be used by succeeding generations of customers. The venture became so successful that tea and cakes at Joe Lyons whilst on shopping trips became an essential part of the English way of life. In 1912 they opened a branch at 175 High Street, Camden Town, (closed 1970), followed several years later in 1923 with another at 299 Kentish Town Road, (closed 1942) sited next to Marks & Spencer department store. Their presence in this road however precedes the tea shop, for Isidore and Montague Gluckstein together with brother-in law Barnett Salmon were partners in Salmon & Gluckstein Ltd tobacco shops and with Joseph Lyons co-founders of the Lyons business empire; the tobacco shop at 197 Kentish Town Road became the site of the Palace Cinema which opened in December 1913. Their inclusion here not entirely nostalgia based but rests on the fact that their catering division were suppliers to the military establishments of vast amounts of fresh meat and meals in all its culinary guises. Among their retail customers were Fortnum & Mason, stocking a variety of their products including sausages. The Lyons food empire eventually encompassed Henry Telfer Ltd., meat pie manufactures later renamed J. Lyons Products Ltd in 1980.

The tea shops and such like never impinged on the livelihood of the pork butcher, they did however face competition from another source the Ham, Beef and Tongue shops, selling as the name implies a selection of cooked items. The number of traders is difficult to asses, presenting an ambiguous classification insofar as trading under that title and recorded as such; in reality many were provision retailers. They arrived hard on the heels of the butchers in Kentish Town Road, among them Henry Roper an early arrival setting up shop at no 3 Old Chapel Row and John Smith at 12 Wolsey Terrace in 1860, increasing to seven in the same line of business; the last Mrs Louisa Beackman, in 1958 albeit by this date titled cooked meat shop.

In contrast to butchers who travelled to markets for their meat, the H.B.T. retailers had suppliers near at hand like T. P. Peacock, with a warehouse in Fortess Road and wholesaler H. James, advertising in the Camden & Kentish Town Gazette. Many of the cooked meat and pie shops kept open to catch the late night revellers emptying from nearby public houses. As Mrs Reeks recalls in the early nineteen hundreds 'There was one shop for cooked food next to Pages butchers, Camden Town, which kept open until mid-night. The front was open and people often stood outside eating saveloys, pease pudding, faggots and pigs trotters'. A later generation of shops which had a reputation as cooked meat retailers, but yet again in reality were provisions and grocery based, would be L.E. Jolly Ltd incorporated in 1921 with shops in Fortess Road, Hampstead Road and High Street Camden Town and further afield. The company registered offices were at 33 Grosvenor Street, until June 1977 the year this firm was dissolved.

MEATHIST MINIATURE

Faggots were also known as Savoury Duck or Poor Man's and reputedly the cause of the Great Fire of London in 1666, having caught fire in the bakers shop. Collard Head is a superior version of Brawn made from pork head meat, which is particularly flavoursome. In the eastern counties of England, Pork Cheese made pork rinds, head, feet and hocks was popular; the meat seasoned and pressed into blocks with the addition of jelly. The curiously named Bath-Chaps are the cheek muscles of the pig when processed as a pork product. The term 'Measly Pork' often found in medieval manuscripts refers to the presence of the cyst stage of (Cysticercus cellulosae) of the tapeworm of man Taenia saginata.

4

MEN, MEAT AND MOTIVATION

The single most important aspect in the preservation of food has been the mechanical application of low temperatures, allowing regular shipments of perishable food thousands of miles across the oceans to arrive at its destination in a wholesome condition. In short to imitate nature and produce cold when and where it is required. The minutia of historical and technical detail, the reader maybe relieved to know, is not within the remit of this book. We may instead draw satisfaction from knowing, whenever we fan ourselves, we can produce cooling without ever understanding the scientific principles involved.

However, with due regard to the men of learning who laboured long and hard to achieve such knowledge, and those business entrepreneurs who applied that knowledge outstanding were: Robert Boyle, Frances Bacon, Stephen Hales, Charles Tellier, Ferdinand Carre, Joseph Coleman, Henry and James Bell, William Thomson, Thomas Sutcliffe Mort, Eugene Nicolle, Gus Swift, Timothy Eastman, Thomas Borthwick, and the Nelson and Vestey brothers. On the lower strata of scientific research enquiring minds were no less intense, George Barth of York Road, James Harrison of Mornington Crescent and Archibald Campbell of Hampstead all lodged patents for various kinds of refrigeration apparatus. The culmination of their efforts and others after many setbacks erupted onto the world food scene in the second half of the nineteenth century.

We may ask why it was necessary to import food and particularly meat. The answer is unequivocal, Great Britain was self sufficient in neither; by this period the average consumption of 120 Ibs meat per head, more than double the average consumption of Europe which required a yearly supply of 600,000 tons. In the beginning the first appreciable meat imports preserved by cold to reach these shores were dressed chilled beef sides from America, using natural ice chambers. This was a new venture for shipping mogul T. C. Eastman, already the largest exporter of live cattle in America. The risks on the transatlantic voyage in October 1875 were considerable, but then again so were the rewards for Timothy Eastman and Gus Swift two of the pioneers involved. These two entrepreneurs had shown the usual Chicago stockyard tenacity and salesmanship, with Gus Swift making numerous Atlantic crossings to clinch the deal with London and Liverpool wholesale importers.

Although impressive as this achievement undoubtedly was, there still remained the millions of cattle and sheep in Australia and New Zealand that were utilized for tinned meat and tallow for want of suitable marine refrigeration; as the time and distance involved transporting meat from south of the equator prevented any possibility of applying the chilling technique. The storage time for chilled beef then being approximately six weeks, the meat stored at a temperature of 29-30 degrees Fahrenheit, so that it does not actually freeze. For the long haul voyages from the colonies freezing was the only option.

When the door leading to success was finally prised open, the vast untapped meat supplies of Australia and New Zealand became accessible. From Australia the steamship "Strathleven" of 2,436 gross tonnage fitted with a Bell-Coleman machine set sail on 2nd November 1879, from Port Jackson near Sidney, for London calling at Melbourne on route. On board in the refrigerated cargo hold were 40 tons of frozen beef and mutton. The 11,500-mile journey took sixty-eight days, arriving at the Port of London on February 2nd 1880. The meat was eagerly awaited on Smithfield Market, and on arrival inspected it was pronounced wholesome and fit to eat. A celebratory dinner of cooked Australian mutton and beef was held aboard the S.S. Strathleven a few days later, and the quality so impressed the principles of the Smithfield wholesalers invited, they signed a statement to that effect, among their number Benjamin Venables of Atholl House, Camden Road.

Within two years, the first ever shipment of frozen meat from New Zealand also arrived aboard the ship 'Dunedin' at East India Docks on the 24th May 1882. The cargo inventory listed 4,460 mutton, 449 lamb, and 22 pig carcasses, all frozen solid and differed from other shipments of frozen meat having been made in a sailing vessel; the journey had taken 98 days, and again using a Bell-Coleman machine the meat kept at 20 degrees below freezing. This together with American, Argentine and later Paraguay established a continual stream of meat imports transforming the British butchers trade forever. There was talk of accolades for those involved, but these would wait another day. The immediate priority was to recoup on the financial investments for themselves and backers, and in 1884 the first advertisement for New Zealand mutton by way of a handbill was circulated in London.

From the outset, marketing strategy relied on high volume and the hard sell to the public. In the early stages, their modus operandi was crude and relentless where price, profit and turnover ruled, renting run down shops in the least desirable locations. The consideration of space and refrigeration was of secondary importance, when the meat became less wholesome it was sold at knockdown prices or consigned to the brine tub to save the worst excesses of waste. Any resemblance to the butcher's craft was purely coincidental, in the early days frozen meat often being sold by the lump. The only advice to shop managers was 'pile it high and sell it quick', and strictly cash transactions only. The frozen beef and mutton of variable quality before the advent of breeding programmes and grading systems.

The freezing of lamb at this early period was not considered a viable commercial prospect, but when introduced later became an instant hit with consumers. The public scepticism of frozen meat, especially in the poorer districts, was soon overcome by its cheapness and widespread availability. Those in the higher income bracket needed more persuading and were doubtful of its merits; their opinions influenced by food gurus Mrs Beeton, Mrs Peel, Mrs De Salis and others. The prejudice centred not on its pallid appearance when thawed, but mainly on associated taste and wetness. A problem recognised by early pioneers in this field and evident by patents taken out detailing processes to thaw the meat before delivery to the wholesalers; such as the British and Colonial Meat Defrosting Syndicate Ltd, in 1902. Advances in modern refrigeration technology have of course minimised the problem. Inevitably the burgeoning domestic refrigeration appliance market expanded rapidly, most prolific G. Kent Refrigeration, the company with manufacturing facilities and sales rooms in High Holborn, London, offering patented ice safes and refrigerators for every conceivable situation from tropical ice boxes to rotary ice cream freezers. Of interest many of the Rotary Knife Cleaners produced by this company can still be viewed in the kitchens of stately homes and from customer testimonials published in 1877 of satisfaction from numerous trades, from this list we are able say one of these hand driven machines was purchased for use by the landlord of the Torriano Arms in Leighton Road, Kentish Town.

HOSTILITY

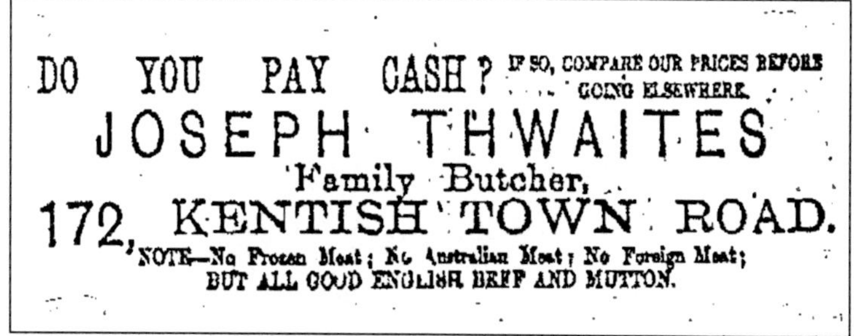

Figure 26: Joseph Thwaits, St Pancras Guardian, November 1890.

It is important to stress here talk of buying frozen meat was tantamount to blasphemy among sections of the meat eating public and retail butchers, who were manifestly against stocking such a new commodity. A 'fiendish diabolical suggestion' wrote one journalist, the 'butchers will never allow their meat to be frozen if they can help it' reported another. From the outset many established retail butchers continued to sell their usual range of fresh meat; at the beginning few if any would stock the frozen, adhering to their anti-frozen and chilled meat stance. Like that of forty-four year old Joseph Thwaites in Kentish Town Road made known using local newspaper advertisements, 'Note: No Frozen Meat, No Australian Meat, No Foreign Meat', his deep seated resentment still obvious ten years after the first imports. It was state a of affairs that signalled serious financial repercussions for the new breed of entrepreneurs and their investors who had colossal capitol sums of money tied up in these ventures. Tough and tenacious, they very soon moved into the retail sector themselves, and there began a rapid expansion of a new and more aggressive style of meat retailer, the multiple butcher. At their inception, perceived to constitute a threat to the livelihood of independently owned butchers, the men behind the multiples were only too fully aware to avoid financial losses, frozen meat by its very nature required speedy and efficient distribution.

WHY NOT, PAY CASH AND GO TO
H. LOVE,
CASH BUTCHER
287, KENTISH TOWN ROAD,
FOR GOOD MEAT AT REASONBLE PRICES.
ALSO AT
SMITHFIELD MEAT MARKET, DALSTON, WANDSWORTH, AND
CALEDONIAN ROAD. NORTHAMPTON

Figure 27: H. Love, St Pancras Guardian, 1890.

The frozen meat trade was by no means the prerogative of multiples, large or small. The plethora of imported meat wholesalers which had grown up around the trade targeted the food retailers with a deluge of advertising propaganda, extolling the virtues of frozen meat and the ease with which it could be marketed. Hoards of salesmen, reminiscent of American carpet-baggers, flooded into every corner of the country offering discounts and trial quantities and promising a quick profit and emphasising how little expertise was required to trade in beef and mutton. The product delivered to your premises or nearest railway station, thereby avoiding the necessity of early morning visits to the market and slaughterhouses.

Figure 28: James Hook, 317 Kentish Town Road, 1903.

With the onset of the industrial age, leading to regular employment and widespread availability of affordable frozen meat, there was a dramatic increase of independent retail butchers, who by virtue of being newcomers without business collateral were often refused credit. This in turn obliged customer payment to be made in cash at the point of sale, where the butcher dealt mainly in frozen meat. Many advertisements to this effect, by way of offering low prices for cash payment and other inducements, can be noted at this period. Better late than never the high class butchers establishments, the last bastions of resistance to frozen meat, gave way in the realization this particular kind of food preservation was here to stay. The truth of the matter was the retail wind of change had been blowing for some time; for Londoners alone consumed the highest proportion of frozen and chilled meat in the United Kingdom, the largest importer of Australian and New Zealand meat.

For the majority of local independent butchers occupying secondary sites in Kentish Town most remained in a small way of business with perhaps one or two branches. Others, typically in the high street, had already acquired additional prime locations elsewhere in London before extending their operation to include Kentish Town.

An example would be Henry Love that as noted in the advertisement, was already in a substantial way of business including Smithfield Market before his arrival here. His tenure in the region of two or three years was of short duration, but long enough to be granted a slaughterhouse licence in the name of family member John Love. The shop continued as retail butchers under the proprietorship of Frederick Turner, and finally Harry Fox and thereafter home to a variety of other trades.

Inevitable rogue elements within the trade resorted to fraud and deception from the beginning. Many customers innocently believing the designation Canterbury lamb meant Canterbury in Kent, the confusion positively encouraged by some untrustworthy butchers, and companies seeking a pecuniary advantage. In addition Australian mutton had a reputation for being of slightly less quality and was at times sold as New Zealand, which in turn was commonly thawed out and sold as English at the higher price to unwary customers. Passing off dubious grades of meat was not representative of the trade in general, and outraged responsible tradesmen like Henry Hooke & Sons of Hampstead Road conducted their business in a fair and open manner. The origins of meat often reflected in the company name like the overtly titled English & Colonial Meat Co, shop at 216 Kentish Town Road.

From the local businessmen with greater enterprise who expanded, and worthy of mention is the prolific butcher and greengrocery firm of H. Ritchie, a possible contender for the highest number of individual shops in one road. He commenced trading in 1898, as a pork & beef butchers beginning an association with Kentish Town approaching eighty years. The first shop at 190 Kentish Town

Figure 29: H. Ritchie, 190 Kentish Town Road, 1903.

Road also provided accommodation for thirty year old Harry Ritchie and his family; one of six premises trading as butchers or greengrocers or both in this road. For readers with an eye for curiosities, what appears to be a large padlock hanging from Ritchie's roller shutter door is actually a milk-can left out each night for the milkman to refill for the sitting tenants or possibly the butcher's tea. A common feature seen in old photographs before glass milk bottles were introduced in the nineteen twenties. There will also be noticed a building reflected in the shop window, which is the Jolly Anglers public house situated opposite on the corner of Anglers Lane.

The landlord at this time being Harry Rolles, possibly one of the Rolles brothers then owners of the Bull & Gate, further along Kentish Town Road. In the years spanning 1950 to 1960, the Ritchie family of retail butchers embarked on a period of expansion, when offices, cutting rooms, meat storage and transport facilities, including petrol pumps, were opened at number one Carrol Place, Highgate Road. The management, office staff, butchers and drivers employed there were responsible for the efficient operation of all the firms' eighteen shops in London, with a suggestion of an additional three shops located on the south east coast area. The continuity of local family businesses like the Ritchie's ensured long term local employment and in many instances a much needed source of living accommodation above the shops. Local resident Geoff Linsay recalls beginning his retail meat career in 1956, at Ritchie's shop in the high street. Whilst in their employ he spent some time working in their Queens Crescent branch, which he distinctly remembers had a very busy trade. The shop at that period was managed by Mr G. Bayliss, with the able assistance of his staff, Eddie, George, Josie and Queenie. As with similar premises in this road selling food, the cellar housed a refrigeration unit and there was no back entry to the shop. Following a period with Ritchie's, young apprentice Geoff moved on to gain a broader experience with others retail butchers before returning to his former employer. He was eventually to manage the Queens Crescent branch where he stayed until the shop ceased trading; his fellow employee of long standing, Chic, lived above the shop.

Figure 30: Ted Ford and assistant, 157 Highgate Road, c1950.

Another contribution comes from Ted Ford who joined Ritchie's on leaving school during the Second World War, and started work in the sausage factory located at the rear of 190 Kentish

Figure 31: Tomkins Butcher, 11 Hampstead Road, CT, 1903.

Town Road. His sausage making skills had to be combined with the process of smoking fish in the oak rooms and shop duties. In the nineteen fifties, while working at a branch shop sited at 157 Highgate Road on the corner of Mansfield Road, the installation of a new refrigerated counter justified maximum publicity and a photograph was taken to mark the occasion. Soon after this event an opportunity arose to relocate to Woking, Surrey, and Ted together with his wife left London. He secured a position with Mathews Butchers Ltd., and remained with them until his retirement. It was around. this time Ritches purchased two shops owned by Hunter & Sons, and a double fronted shop owned by G. F. Kimber all sited in Kentish Town Road and when these were added to their existing outlets made Ritchie the leading meat retailer in this road; an advantage that buttressed the firm against the changeover from predominately high street shopping to one of supermarkets. In April 1994, after nearly a century of service J. H. Ritchie closed its doors on what had been a successful business largely due to the personality of one man.

Others were inclined to pool their resources like the partnership of Messrs Philip and Tompkins in 1885 selling New Zealand mutton in bulk from shops in Marylebone, Goodge Street and Pimlico. This was a widely used standard business arrangement but one of high failure rates with the two participants eventually going their separate ways, and the creation of Daniel Tomkins & Co.

The Tompkins shop in Hampstead Road, one of eight in the new business setup, dates from 1892 when Edward Tomkins eldest of three sons, the others Frederick and Edward, is recorded as the named registered licensee holder No 267 of the slaughterhouse at the premise; the previous owner and licence holder being Joseph Bazzoni. The business titled changing by 1915, on incorporation as a limited company and registered as Tomkins Bros Ltd. One of whom, living above the shop, was Harold aged 31 the son of 56 year old Edward Tomkins and both originally from Bexley Heath, Kent. His father and mother Elizabeth aged 57, living at branch shop number 5 High

Street, St, Marylebone. By 1940 only two other shops remained at numbers 41& 61 Great Titchfield Street until finally the Hampstead Road shop survived into the late nineteen–sixties, listed as number 52 under the sole proprietorship of Frederick George Tompkins.

BULL BROTHERS,
Butchers.
173 HAMPSTEAD ROAD, N.W.
Families supplied with meat of the finest quality on the lowest terms.

Figure 32: Bull Brothers, Christ Church Magazine, 1901.

In the same road Frederick Bull of whom little is known, other than in 1926 a butcher named William Bull appeared at Marylebone Magistrates Court to answer charges of keeping his shop premises at 24 Upper Park Place, Dorset Square, Marylebone as a betting house. The arresting police officer gave evidence the defendant was in a good way of business as a butcher and had also been carrying on an extensive betting business for three years. Found guilty, William Bull was fined £50 pounds with costs or 51 days imprisonment. The Oldfield Brothers shop in 1915 was under the proprietorship of Norfolk born John Oldfield. They are believed to be descendants of Samuel Oldfield in 1861 with a shop in Torriano Avenue, Kentish Town.

THE BIG PLAYERS

With the twentieth century came the new age of the multiple meat trading companies, the big players that by the nineteen twenties were household names reaching out to every high street, controlling approximately 4,500 outlets. For many the rise to business prominence as a direct result of the founder's entrepreneurial spirit by early participation in every facet of chilled and frozen meat from retail to imports. Always the subject of business cartels and mergers within the trade, sold, resold, renamed and revamped, until finally falling out of step with changing trends and facing the inevitable consequences. Among the early exponents John Bell & Sons, of London, Glasgow and Liverpool, established as retail butchers in 1827 and incorporated in 1888. Three years later sons James and Henry joined forces with Timothy and Joseph Eastman, the American chilled beef and cattle exporter previously mentioned. The two concerns traded as Eastman's Ltd, their combined nationwide outlets totalled 1,400, the second largest chain of butchers.

OLDFIELD BROTHERS.
Family Butchers,
81 OSNABURGH STREET, N.W.
Families waited upon Daily for Orders.

Figure 33: Oldfield Brothers, 1901.

Some business ventures travelled a different route, the first rung of the retail ladder in 1880 for Messrs George E. Lowe and Herbert Lea was a stall in Tamworth market, Staffordshire. They began by selling American chilled beef, expanding the range to include New Zealand, Australian and Argentine frozen meat. In 1892 premises were rented in Gold Street, Northampton, the average weekly takings amounting to £70 pounds. The business accelerated rapidly and in 1896, incorporated under the title London Central Meat Co. Ltd., and in the preceding years of the First World War they had amassed over 500 branches. Another war and another generation of shoppers would know them better as Baxters (Butchers) Ltd. Away from the south was multiple W. & R. Fletcher Ltd., whose sphere of operations was predominantly in the North of England.

The Liverpool based family firm of James Nelson & Son Ltd, founded in the 1840s began as cattle salesmen. Later as successful pioneers of the frozen meat trade, quickly building a global business empire that encompassed not only meat production but cold storage and shipping, with offices in all the major meat trading countries. In 1885, they opened their first shop, becoming one of the leading exponents of what was then a new concept in meat retailing. By 1910, shop numbers totalled over 1,500 countrywide, with 250 of these in London and surrounding suburbs, one at 190a High Street, Camden Town in 1904 and another sited at 62 Seven Sisters Road, Holloway.

Figure 34: Eastmans, Advertisement Card, c1912.

A continuing feature of some businesses however is takeovers and mergers, the most prolific predator, the Vestey Group Ltd, formally the Vestey Brothers now Unilever plc; whose commercial appetite resulted in acquiring many of the premier retail and wholesale meat companies. Any attempt to relate the full or even partial company history within the page limits of this book a futile exercise; suffice to say such was the enormity of their business empire from raw materials to food shop companies in 1911 they founded the Blue Star Line shipping line to facilitate trade with the continents of the world. Their list of business acquisitions alone running into hundreds of pages that included Eastmans, W & R Fletcher, James Nelson & Son Ltd, in addition creating their own meat retailers the British & Argentine Meat Co Ltd. Thereby assuming an unassailable position as the leading meat retailers in the country, a position unchallenged on the domestic retail meat trade until after the Second World War when the food industry conglomerate Fitch Lovell Ltd first entered meat retailing.

The company originally formed from a merger between two long established and distinguished firms Fitch & Son Ltd and Lovell & Christmas Ltd. Their venture into meat retailing began in 1950, which in hindsight had limited long term prospects. However, in that year was bought seven shops in London trading as Layton & Burkett, to form a nucleus of shops from which to expand. The Burkett side of the partnership of long standing had several shops; one sited at 77 Carlton Road in the eighteen nineties the other in Kentish Town high street. There were further acquisitions of businesses in quick succession which by 1958 had increased the number of Layton & Burkett shops to forty-nine. One year later Fitch Lovell, in a share exchange deal, gained control of West Butchers Ltd a public company with eighty branches. Following this latest acquisition West butchers were merged with Layton & Burkett to form a new meat retailing company titled West Layton Ltd.

The West butchers enterprise was the idea of butcher's son James Preston, the second eldest of three brothers all born in St Pancras and living in Charlton Road, Somers Town. On the death of his father Henry Preston, eldest son James, with brothers Joseph and Frances inherited the shop. A man of outstanding ability, by the early nineteen hundreds the business with a number of shops and office on Smithfield market had become incorporated.

A flamboyant character and fastidious in dress and manner, he arrived at market in his own carriage each day, after the mornings business he would visit his shops one of which was at 82 Queens Crescent and another in the high street Kentish Town Road.

In 1918 James dissolved his original company and formed a new limited private company using the title West Butchers Ltd., the title name taken from the location of the Preston company head office in West Smithfield. In 1934, the business became a public company and James Preston subsequently retired. By 1979 West Layton had grown to 170 shops, with branches throughout London and the Home Counties. A company livery of green and gold was chosen and in addition the introduction of delicatessen counters. This service in some areas, proved so popular the larger shops were revamped and renamed 'West Continental'. The previous occupants at their branch at 222 Kentish Town Road consisting of James Preston, West Butchers Ltd and West Layton almost mirrored the rise of Fitch Lovell Ltd as a major force in the retail butchers sector; their last important foray into the retail business market in 1960, being the purchase of R. Gunner Ltd and J. Richards. The Joseph Richards concern, with thirty branches, predominately operating in the south of London.

Figure 35: London Central Meat Co Ltd, c1930.

The original Gunner enterprise was considerably more diverse and the business extensive. The founder Robert Gunner began trading as a cheesemonger c1903 at 138 Essex Road, Islington, and the family lived above the shop where a son Ernest Gunner was born. Business increased by adding ham and hogmeats to boost sales even further and in a relatively short period of time Robert had opened other shops in the same road and in nearby Chapel Street market and at London Fields and Cricklewood. By 1914 a warehouse and offices had been rented at 28 Cowcross Street Smithfield. At this juncture his trade had become one of provisions retailer selling considerable quantities of bacon and it is reputed he placed the first order for bacon from the newly formed Danish Bacon Company founded by Charles Hansen.

It was not until the First World War that Gunners entered the retail meat trade, when a butchers shop was purchased from German pork butcher assumed to be Alfred Wenninger 11 Broadway, London Fields. Thus Robert Gunner first entered the retail meat trade and thereafter continued to expand both meat and provisions retailing operations.

In 1928 business operations were expanded still further by building a new factory and warehouse at 111-115 Lever Street, EC1, and in the early 1930s they also opened their own bacon producing factory at Letchworth. Following the death of Robert Gunner his son Ernest Gunner, already active in the family business, decided to concentrate his efforts on his bacon manufacturing interest in Canada. He subsequently passed control of the 118 Gunner shops by arrangement to Joe Richards, proprietor of J. Richards (Butchers) Ltd. Thereafter R. Gunner Ltd passed into the ownership of Fitch Lovell Company, who themselves eventually became part of the Unilever Empire.

Returning to the advent of frozen and chilled meat, competition came from all quarters, even some multiple grocers unable to resist the financial temptation of adding this new commodity to there range of goods. As the trickle of frozen meat from the 1880s turned into a flood, the lure of this untapped source of revenue for Sainsburys proved irresistible and by 1910 New Zealand meat was added to their range. A familiar sight to every shopper, the expansion of these multiple grocers had been staggering; of the major players on the outbreak of the Second World War, each had the following number of shops: Maypole Dairies 997, Home & Colonial 798, Sainsburys 255 and Liptons 449, the last named established in 1871 by Thomas Lipton in Glasgow. It may be noted Sainsburys were the smallest of the group. Other small to medium sized grocery companies such as L E. Jolly Ltd and Pearks Stores owned by Meadow Dairy Co Ltd emerged later on. Any discussion of national figures would be arbitrary, but census information reveals by 1911 there were 196 shops of all kinds per 10,000 of the population in England and Wales.

Figure 36: Thomas Burkett & Co, 249 Kentish Town Road, 1903.

As yet unmentioned there emerged one other competitor, the Co-operative Society operating under a very different business philosophy. The movement which is accredited to social reformer Robert Owen (1771-1856), one of his better known disciples George J. Holyoake is buried in Highgate cemetery. Robert passionately believed that only in the spirit and practice of cooperation and community could the evils that beset the country be defeated; he instituted social festivals for exchange of ideas and neighbourliness with equitable labour exchanges for the barter of products produced, the value of each being adjudicated by a committee and a ticket issued. One of the centres in London, The Exchange and Co-operative Bazaar, was located in Gray's Inn Road, near Kings Cross, not far from where Robert Owen lived for a time at 4 Crescent Place now Burton Place. Of interest in 1874 the Midland Cooperative Store is noted in King Street; now Plender Street) Camden Town.

It was the Rochdale Pioneers (Rochdale Cooperative Society) of 1844 that decided a dividend should be paid, and with this decision came into the market place shopping, sharing and socialism by 1890s, extending into almost every area of their business and commerce. The background, insofar as it applies to the meat trade, is believed to have commenced in Woolwich when a co-operative meat shop opened there in 1805 closing six years later. In 1846 Rochdale Co-operative Society, bastion of the movement, placed a weekly order for 100 lbs of beef and mutton from a local butcher and in 1851 Oldham purchased 'half a cow and a sheep', all for resale to its members.

Other early attempts to trade in meat were all of short duration and are recorded for Sheerness Co-op Society in Kent, 1860, and Failsworth, Manchester in 1872. Inexplicably when it came to meat merchandising early co-ops were singly unsuccessful or uninterested, however by the 1890s co-operative meat sales had increased considerably and by the nineteen twenties from a total of 1,314 co-operative societies trading, 599 were dealing in meat. For all practical purposes the handling and selling of meat virtually undistinguishable from the average butchers shop. It was during this period the movement spread from their traditional northern heartland down into the south and London. Until rationalisation innumerable district titles were used e.g. Ilford, Enfield, Edmonton, and Woolwich causing considerable overlapping. The Edmonton Co-op for instance had a double fronted shop at 122 Kentish Town Road on the corner of Rochester Road, until absorbed into the London Co-operative Society Ltd. The London Co-op was formed in September 1920, by the amalgamation of the Edmonton and Stratford Societies and began a period of consolidation ending in 1938 of all the Societies in the Greater London area. By 1952, with over 550 establishments varying from department stores to corner shops, the LCS covered almost the entire range of retailing services in addition to manufacturing and processing, dairy and stock farms, abattoirs and the Cooperative Wholesale Society to supply the shops and stores. A further reorganisation in 1981 amalgamated the LCS with the Co-operative Society.

J. H. DEWHURST LTD

These then were representative of the multiple retail butchers, their roll call of names once familiar to every shopper in the land are hardly remembered now with the exception of J. H. Dewhurst Ltd, this most famous of retail butchers became the biggest player of them all. Yet despite being a household name very few customers had knowledge of their origins, or indeed the chance decision that catapulted the name into the public arena. A decision arrived at when Ronald Vestey the grandson of Samuel and founder of the company, that later became the Vestey Group Ltd, decided to bring all his retail meat division shops under one name. He chose at random the business formerly carried on by meat purveyor and pork butcher John Henry Dewhurst of Southport in Lancashire, a moderately sized operation acquired by the Vestey family as a going concern in 1915.

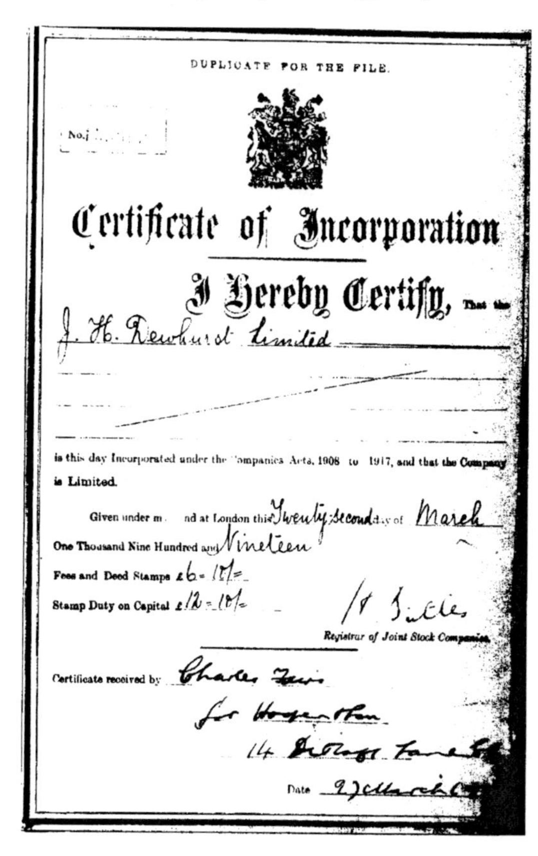

DUPLICATE FOR THE FILE.

No.

Certificate of Incorporation

I Hereby Certify, That the

J. H. Dewhurst Limited

is this day Incorporated under the Companies Acts, 1908 to 1917, and that the Company is Limited.

Given under my hand at London this Twenty-second day of March

One Thousand Nine Hundred and Nineteen

Fees and Deed Stamps £6-10/-

Stamp Duty on Capital £12-10/-

Registrar of Joint Stock Companies.

Certificate received by

Date

Figure 37: J. H. Dewhurst Ltd, Certificate of Incorporation, 1919.

A furore ensued among his senior management, with argument and passionate pleas for their respective meat company name within the group to be considered. However eventually dented egos and disappointments were soothed, and a compromise was reached. Although in reality strictly business considerations dictated the restructure and retention of Fletchers Ltd in the North for imported frozen meat, Eastmans Ltd for predominantly fresh meat serving the West Country and Kingstons Ltd for robust market type shops. Their interests in Scotland were served by butchers Alex Monroe Ltd and the pork and delicatessen retail trade was represented by T. W. Downs that later became the nucleus for Downsway supermarkets. This first round of restructuring and consolidation did not mean an end to future expansion and when the opportunity presented itself to purchase another reputable company, the highly respected retail butchers Hammetts of London were added to the portfolio.

In 1969 with a new progressive chairman, Collin Cullimore, at the helm and confident in the future; the company embarked on a programme of modernisation coupled with sustained high profile advertising. This included the gradual replacing of the old style title with 'Dewhurst the Master Butcher' for all new and revamped shops, with the exception of Scotland under the Alex Monroe banner and a few prestigious Eastman shops.

The core of any business is the quality of professionalism, and in 1972 a nationwide chain of staff training schools were opened. This, together with widespread shop locations and a host of other innovative ideas such as 'Cooking with Dewhurst' books and Food & Farming educational resource packs for schools, ensured the company name remained synonymous with the retail meat sector. Confident in the future construction of a prestigious building to be named Dewhurst House in west Smithfield was completed in 1975, and along the way adding sixty-six Freezer Fare Stores to the Vestey group.

In 1980, the Dewhurst reputation became tarnished amid accusations of tax evasion and other matters following a seven month investigation into the affairs of the Vestey shipping and meat companies by Phillip Knightley, a Sunday Times journalist. In December that year the Vestey owned Freezer Fare Stores were sold to the Argyll Food group, however J H. Dewhurst Ltd continued to trade successfully, claiming in 1989/90 to have 1,260 shops. Whether a sign of foreboding or coincidence, their branch at 149 Camden High Street closed, two years later when the lease was surrendered to the landlords, Marks & Spencer. At local level the company began their presence at 313 Kentish Town Road in the early nineteen-thirties with a branch on the western side and ended their reign across the road at number 222 some sixty years later, whilst nearby at 151 Fortess Road they occupied the site until July 1984. There were no Dewhurst branches opened in either Malden Road or Queens Crescent, although a branch is recorded at Haverstock Hill, Hampstead. In Camden Town they purchased property in Arlington Road to provide hostel accommodation for trainees from across the country, to overcome a severe shortage of experienced young butchers in London.

However in March 1995 the unthinkable happened and the company went into receivership, with Messrs Ernst & Young appointed receivers. The Dewhurst name survived, following a

Figure 38: West Layton, company Logo, 1968.

management buy-out in August of that year with the new team undertaking refurbishment of shops and updating equipment, in addition to a more innovative approach to modern day customer meat requirements. The public outwardly noticed no appreciable difference albeit a subtle change had been affected with the deletion of the initials J. H. from the original title. And despite problems that adversely affected the whole meat trade, the new company with the old name continued to trade with two hundred branches, twelve concessions, mobile butchers shops and a staff of 1,500. Ironically with only a fraction of the shops owned by their predecessors, they still retained the position as the largest multiple retail butchers in the country. An indication of how drastic the national decline in retail butchers' shops had been. The climax about to be reached in March 2005, when the company was purchased by the Devon based Lloyd Maunder group, operating seven butcher's shops, hatcheries, farms and feed mills. At the time of sale Dewhurst reduced to operating 109 shops and 500 staff. One year later in March 2006, almost to the day the Lloyd Maunder Group called in the receivers citing rent increases, energy prices and poor trading conditions the final demise of Dewhurst butchers seen by many in the trade as the defining moment in the history of the retail butchers shops in our high streets.

The Hammetts business was founded in 1901 by twenty-one year old Richard Christmas Hammett and Partner, trading in meat and produce. He was born on Christmas day 1880 at the family home in Devon, and later moved to Battersea nearer his premises situated in Farringdon close to Smithfield Market and the Butchers Hall wherein he soon rose to prominence. He was co-opted to serve on various food supply and distribution committees during both world wars and was a driving force for education in the meat trade; a belief that culminated with the establishment of the Institute of Meat. He also found time to publish text books for butchers of which he was co-author. Always deeply involved in City of London affairs he served as common councillor and also served on the London County Council. The years 1947-49 proved exceptionally rewarding, appointed a Sheriff of the City of London and twice as Master of the Butchers Company. The business was incorporated on the 8th December 1925, as butchers, poulterers and graziers, with head offices at 14 West Smithfield, London, EC1. The company had branches in most areas of London including Finchley Road, Hornsey Road and in the 1930s claimed to be the largest meat retailer in London. The company retail shops were eventually acquired by the Vestey Group. Mr R C. Hammett died in December 1952 and a memorial service was held at the Priory Church of St Bartholomew, attended by the Lord Mayor of London and a host of dignitaries. His name survived in R C Hammett Ltd Wholesale Meat & Meat Products, Northants.

Figure 39: R. C. Hammett Ltd, 1939.

CATERING BUTCHERS

The following supplied meat and poultry to Hampstead Work House as recorded in the account books for 1734-1739: poulterer Thomas Daniel with butchers Thomas Davis, Cleaver and Lucas. The first mentioned poulterer Daniel and butcher Davis both have single entries in Pigots London and Provincial Directory 1823/24, for Kentish Town and Hampstead sections respectively. Of the last two named, these butchers appear in the guild records in the Apprentice Rolls for 1585-9; the dates would seem to disqualify these two particular butchers as the suppliers, albeit there is a possibility their descendents are the named butchers in the account books. With regard to Lucas a proprietor of a butchers stall with that name; Richard Lucas traded in Leadenhall Market, in 1775. In that year on the 17th November, several cuts of unwholesome meat were confiscated from the Lucas stall and he was summoned to appear before the Quarter Sessions.

From the beginning of the eighteen hundreds in common with other authorities; St Pancras Vestry, then located in William Street regularly advertised in newspapers for all types of businesses from, chandlers to printers, able and willing to enter into contracts to supply the needs of the workhouse. In one dated 31st August 1829, tenders were invited for bread, meat, butter, cheese, soap, candles and small beer, i.e. mild beer. At this period, these transactions very much conducted on an ad-hoc basis with scant regard for accountability with the inevitable consequence leading to well publicised court cases and public concern from which grew a greater awareness of unscrupulous suppliers and a more sympathetic understanding of the inmates needs. Institutions like Middlesex prisons at the houses of correction situate at Coldbuth Fields and Westminster Vestry and the House of Detention, Clerkenwell, began detailing their precise requirements, particularly food and especially meat.

"Wanted: Good ox beef without bones, consisting of the clods and stickings, ox head on average weighing 25 lb each, shins and legs of beef, mutton consisting of breast, neck and ribs, and mutton from other parts of the carcass for use in the infirmaries".

Figure 40: Mills Bros, 58 Osnaburgh Street, NW1, 1890.

The vestry officials becoming ever more vigilant and the commodities and services required decidedly more varied. With the arrival of frozen meat many suppliers switched to contract butchery, supplying hotels, restaurants, public houses and other establishments. The following view on the subject is from an unknown hand, but nonetheless was pertinent in its day:

'The contractor is patron of the frozen meat trade to some purpose; he supplies frozen meat in considerable bulk to unions, asylums, and other institutions in England, and some Smithfield traders made a speciality of catering for this kind of trade. A saving of £1000 pounds a year is easily effected in the large institutions by substituting frozen meat for the home-produced variety. There were protests from the inmates when such steps were taken, what is good enough for peers and the beef steak club will not do for the paupers, or could it be that officious guardians arouse the grumble of old fashioned prejudice'.

In the Regents Park area a mention of G. E. Orlick & Co, 180-184 Albany Street a well known meat wholesaler's at the corner of Redhill Street near the Crown and Anchor public house, and an interesting advertisement from wholesaler and retailer the Mills Bros that in common with most butchery businesses jealousy guarded their reputation: the business was incorporated soon after the date of this advertisement, as far as is known there were no home grown Kentish Town wholesale and/ or catering butchers with a nationwide reputation. All were in the lower ranks like Mr G. Cawdry of R & C Butchers who began the catering side of the business in the nineteen sixties operating from his retail shop at 58 Queens Crescent. The catering side eventually needed larger premises, and was transferred to 122 Kentish Town Road, a former London Co-operative Society shop opposite Dunn's Hat and Clothing offices. The business continued to prosper, however economic events in the 1980s effected businesses weak and strong, and both the shop and wholesaling business ceased trading. The origin of R & C butchers was a business partnership formed between Messrs Anthony Ross and George Cawdry, which included shops in the Hackney and Bow area. Experiencing a similar misfortune in this type of business during the 1970s Mr Leslie D. Aston, proprietor of Martins Catering Butchers, 65 Torriano Avenue, in Kentish Town.

Figure 41: Horace Tipple, 1923.

These were the local providers and far above their capacity the catering firms that exist today that supply airlines with ready to eat meals to feed the millions of passengers that crisscross the globe. One of many I.M.S. of Smithfield the name belying the company is also located on the industrial estate in Cedar Way, off Camely Street. The business incorporated on 20th March 1970 and operating from modern temperature controlled buildings formally used by Sparks Catering.

Figure 42: Pike, Biggs & Fairfax, Logo 1989.

In some instances contract butchers have progressed from simply supplying meat and offal to multi million pound industries, tailored to the needs of the catering trades, institutions and supermarkets. One, as many Kentish Town residents in the vicinity will be aware, is Fairfax Meadow factory at 24/27 Regis Road; located on a great swath of land vacated by the Midland Railway opposite the underground station. The company which styles itself the 'Exceptionally Fine Butcher' is the largest specialist catering butchers in the United Kingdom. It produces millions of portion controlled products, and handles tons of meat per year and has a fully equipped technical laboratory. Holders of two Royal Warrants, the company has head offices in Derby with depots nationwide.

Their beginnings are related by former director Henry Tattersall, 'A Mr G.W. Biggs purchased a shop in Greenford in 1939, and this grew by 1961 to eight retail shops and two catering establishments. I joined G. W. Biggs at his Greenford shop in 1955; our first catering clients were Lyons Tea factory nearby. The firm continued to expand, achieving a one million per annum turnover by 1967. In April 1969 I purchased a shell of a factory at 188 York Way, Kings Cross, to convert into a catering and manufacturing unit. The factory employed 150 people, and because of staffing difficulties the retail shops closed. The catering side prospered and our customer base expanded, supplying such customers as Berni Inns, later the Great American Disaster and renamed the Hard Rock.

In December 1981 G.W. Biggs (Harrow) formed a partnership with W.J. Pike from Smithfield Market, both George and Bill Pike had retired some years earlier. Using the title Pyke and Biggs plc, this proved a highly successful venture with net profits in 1984 shown as £1.2 million'. Mr Henry Tattersall retired in 1985, with the satisfied comment that Lyons Tea factory were still clients when he left.' During this period Fairfax became involved and the name changed to Pyke, Biggs & Fairfax, the Fairfax part of the title taken from Fairfax Road, Swiss Cottage where a former partner lived. They eventually became part of Hillsdown Holdings, the Buxted Poultry to Smedley's food group. The trading name changed yet again to Fairfax Meadow plc, the latter title taken from Meadow Farm Produce, in the Hillsdown group portfolio. By 1990 the company had reached an annual turnover of 18 million.

FISHMONGERS

The business life span of fishmongers in London numbering approximately seven-hundred in 1895; many also selling poultry and game were indelibly linked with their cousins in the meat trade and consequently suffered the same downturn in fortune. Slater (Fishmongers) Ltd with twenty retail outlets by 1912, one of them at 58 Belsize Lane, the other at 95 Haverstock Hill nearby the Sir Richard Steel public house on the fringes of West Kentish Town. Sir Richard, essayist and co-editor of the Tatler and the Spectator periodicals, spent periods secluded in his cottage when creditors became too persistent, his hideaway 'Steele's Cottage' sited near where the pub now stands. The cottage was demolished in 1867 to be replaced by the road which bears his name. Fishmonger's wife Mrs Fanny Gaubert would have no such enduring recognition, save in her lifetime two large hand painted sign boards fixed above her husbands fishmongers shop near Castle Place at 133 Kentish Town Road, and photographed in 1903 together with three single-story shops; a pictorial record emphasising the historical importance of photography, the building a surviving example of 18th century architecture. It is also interesting to reflect the shop of Boot

Figure 43: S. Gordon, 1996.

and Shoe maker John Mason next door appears in a photograph over seventy years later, plus a restaurant and fish bar dating from the 1880s. Also worthy of mention Samuel Gordon trading in fish and meat, who opened for business in 1808 at 76 Marchmount Street, Tavistock Square; the shop still in situ and bearing his name still advertising for business in 17th October 1996 issue of the Camden New Journal.

The Dewhurst of fishmongers and poulterers were Mac-Fisheries, a lock stock and barrel organisation from trawlers to high street shops, with tremendous financial resources and blue chip credentials. The company was founded in February 1919 and originally titled 'The Island Fisheries', a name quickly discarded and registered as Mac-Fisheries Limited. The first of 400 branches nationwide opened in the same month at Hill Street, Richmond on Thames. A commemorative plaque was originally attached to the facade of the shop. Their retail division included shops at Hampstead, Holloway, Camden Town and other areas of London but not Kentish Town that for some inexplicable reason had to content itself with a returned empties depot at number 1 Malden Place opposite the Court picture house originally named Gem Picture Hall c1910, and possibly the first cinema to open in Kentish Town. The area lost a popular source of entertainment in 1958 with the final closure of the picture house.

The Mac Fisheries enterprise was the genesis of an idea by the first Lord Leverhulme (William Heaketh) who owned the islands of Lewis and Harris in the Outer Hebrides. Conscious of the long-term well being of both the Islanders and the local fishing industry, he decided the most constructive help would be to provide a new outlet for their catch. A private company, with Leverhulme as chairman, was formed with the declared aims to establish and control multiple fish shops throughout the country. Having embarked on the venture he realised the islands could not supply all the needs for the potential number of shops envisaged.

Figure 44: Mac Fisheries Ltd, Logo 1960.

On 28th March 1922 Mac-Fisheries following a rapid programme of acquisitions and expansion became an associated company of Lever Brothers Ltd. Everything about the new company oozed professionalism and confidence even down to the registered trademark, the work of George Grey a respected figure in the field of Heraldic design. His design comprising a circle surrounding a milled rule, the St Andrews Cross and four fishes and incorporating 'For All Fish' the company motto. Maximum advantage was made of publicity, and to this end a high profile shop was opened in New Bond Street. Here the company mounted impressive displays of fish, calculated to attract the attention of press and pedestrian. Advertising campaigns 'Watch Mac-Fisheries' and "Best Fish Value', based on price, complete with a wavy line logo that appeared in the national press from the 1920s onwards, together with cookery recipes. In time the company would irrevocably change the style of fish merchandising. From the beginning, architect Leslie Mansfield set about creating a standardised and dignified shop appearance using colours of blue and grey. Out went the odour ridden and unhygienic image of fishmongers with newspaper wrappings; in came a strict adherence to cleanliness, with ceramic tiles and refrigerated counters as the latter became available.

Following post war de-control, trading conditions returned to normality with Mac-Fisheries shops restating their position as the leading fish retailer. The underlying trend in the retail food sector however was one of change. Inroads by supermarkets and a greater reliance on cheaper pre-packaged convenience foods; quick frozen fish fingers, cakes and burgers etc, eroded the traditional fish or meat meal. With the arrival of fast food chains the situation became irreversible, and in 1985 the parent company Unilever sold Mac-Fisheries to Clipper Sea Foods Ltd.

Of the 1st Lord Leverhulme, his contribution to the commercial wealth of the nation remains anonymous to the public at large, his early enterprises hardly remembered. For visionary men of his time were the rule rather than the exception. But as resident of 'The Hill', Hampstead Heath there is much locally to commend him, the house situated in North End Road, a short distance beyond Jack Straws Castle public house. 'The Hill or Hill House, its previous name, has been described as the most beautiful house on the heath. Leverhulme purchased the property formerly belonging to descendants of Samuel Hoare, a city banker, in 1904. It is a sobering thought that history records in this house, many are the negotiations and business agreements breathtaking in scope that effected world trade and thereby millions of people.

Un-associated with Mac-Fisheries in size and scope on the south side of Crogsland Road at 54 to 58 opposite Kirkwood Place which formed part of Haverstock Hill school playground, were W. & F. Fish Products. They specialised in curing and smoking fish including bottling pickled herring and onions. The factory afforded local people the opportunity of working from home peeling onions, mainly women who could be seen going to and fro with prams full of onions. The work provided a much needed source of income often referred to sneeringly as pin money, meaning for hair perms or luxuries. Contrary to the cynical views the small amount of money received was vital for their subsistence.

MEATHIST MINIATURE

From the first imports of New Zealand frozen meat from 1882 to 1910, seventy-two million frozen carcasses, mostly sheep were exported to Great Britain. Ever the entrepreneur, in recognition of his frozen meat empire, William Vestey, the first Lord Vestey chose to include an iceberg on the family crest. To disguise the monotony of eating mutton at every meal and yet again at Christmastide, Australian pioneers in the 1880's coined the phrase the 'Colonial Goose' for legs of mutton served on this special occasion.

5

ROYAL PATRONAGE

Always professionally class conscious, with the importation of frozen meat in the 1880's, categories or type of trade became more defined. The two main classes, either working or high class, were the master butcher who slaughtered his animals and the meat purveyor who purchased direct from the wholesale markets. The working class trade, known as colonial or cutting butchers and the high class butchers who by definition aspired to serve the more affluent members of society and obviously a prerequisite must be a location where a preponderance of such people lived. Its origins emanate from the 11th century when guild merchants consisted mainly of the burgesses and other elite persons of the town. The Burgess as shop owners and members of the butcher's guild were considered men of good standing as opposed to stallholders within the community and tended to attract the higher class of trade. The pecking order judged by their peers based on shop location, quality of meat sold and to whom, re-enforced by the social class system and good old English trade snobbery. Top honours in this category at the turn of the 19th century, we may irreverently ascribe to royalty's neighbourhood purveyor Hall & Sons, 21 Buckingham Palace Road, surely the most enviable butchers address in the realm, in 1904 serving H.M. King Edward V11 and H.M. Queen Alexandra.

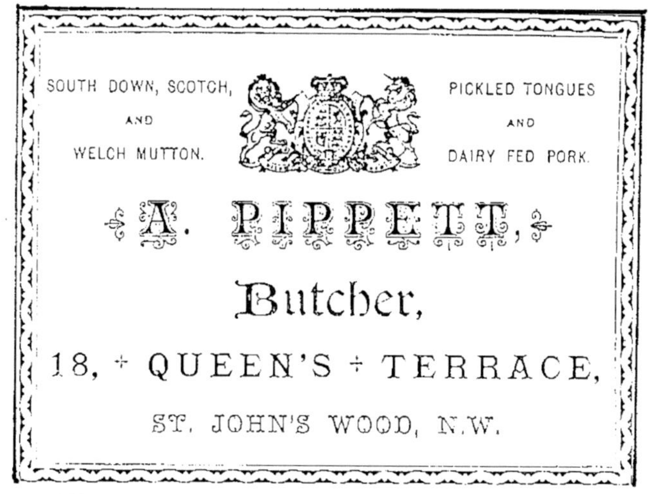

Figure 45: A. Pippett, Queen's Terrace, 1885.

Should the reader doubt the authenticity of the introductory paragraph the following guidance appeared in 'A Practicable Treatise by Specialist in the Meat Trade', published in 1929. 'Usually if a working class population deal in a shop the professional classes will go elsewhere. This does not mean the wife of the professional man is a snob, but she demands a different type of service from that of the wife of a labouring person. The latter will oblige by carrying the goods home herself, whereas the former prefers they should be sent. The district under review may be entirely working class in character with the bulk of the breadwinners doing hard manual work, with a small sprinkling from the clerical profession. Conversely the population may be largely clerical, but still working class'. An echo of the class-divide once prevalent; a belief passionately endorsed and accentuated by the Victorian's adulation for royalty and the upper classes.

Then as now, most keenly sought by merchants and traders alike, was that most coveted of all accolades, Royal Patronage, the use of such accolades in all its guises, Thomas Jefferson (1743-1826) liberal statesman and third president of the United States, had earlier called 'a very great vanity' that America did not intend to copy. A more historically accurate term would have been 'The King's Takers', as purveyors were once called. In feudal times it meant the absolute right for purveyors appointed by the reigning monarch to procure goods for the Royal household without payment; this was later amended to compulsory purchase, often at prices fixed by them and resold on their own behalf.

The custom of purveyance was held in contempt and fear by all who had the misfortune to suffer its consequences. As Archbishop Islip, from whence Islip Street takes its name as a former part of the Christ Church Oxford land tells us in the reign of Edward III. 'Every old woman trembled for her poultry until the King had passed'; the custom was finally abolished by Charles II. The granting of a royal seal of approval certainly put a meat trader above his contemporary's, and the royal household were par excellence when it came to filling the royal banqueting tables and larders.

From the traders view point the object of the exercise was to increase business, and nothing was more calculated to achieve this aim than advertising the fact you were a Royal Warrant holder. In the advertising examples shown, the use of so many styles of lettering in one advertisement maybe thought excessive; however the royal insignia confirms a solid dependable businessman. The first Monarch to allow tradesmen to use the Royal Arms was King William IV. Although it seems there is no record of an independent proprietor based in Kentish Town that has held a royal warrant for a retail butchers shop. From necessity therefore, we must make an excursion into the neighbouring districts beginning with the business under the proprietorship of Mrs Anne Pippett at the time her advertisement appeared in the 1885-6 Hampstead and Highgate Directory; at which period her son Samuel was affectively in control aided by other members of the family.

Figure 46: Barrett, Englands Lane, 1885.

By 1901 it is believed Samuel had married and the couple had a young child named Roger aged five years old. Previous to this the business had been established by George Pippett, who in the eighteen-fifties advertised a house and shop to let in St John's Wood suited for fishmongers or any other business; with the rejoinder "Any person with a little capital and industrious habits may depend upon success" perhaps mindful of his own high standing in the local business community. He was obviously public spirited, having some years later advertised "Dog found on Hampstead Heath, will owner please apply with description", the outcome unknown. He also placed numerous advertisements during a ten year period in the classified section of newspapers for cooks and housemaids. A time of high employment opportunities for domestic staff, which no doubt caused the situation.

Moving on to the Barrett family at 40 England's Lane and other locations. The grandson, and last member of the family to occupy the shop of the founder, the late Frederick F. Barrett began his career with the Zwanenberg company meat wholesalers on Smithfield market. His grandfather started the business in the 1860's, and Frederick took over upon his father's death in 1935. A Master of the Worshipful Company of Butchers in 1974-75 and a life-long resident of Hampstead, he died in 1980. A familiar name in the local history of meat retailing, recorded continuously from 1855, beginning with Thomas Barrett proprietor of number 13 Grange Road, Kentish Town. Other shop locations under the same surname include Flask Walk, Southill Park, Crowndale Road, High Street Camden Town, Goldington Street, Somers Town and Kilburn. However their relationship, one to another cannot be determined with any degree of certainty.

RANDALL BROTHERS

Of equal prominence Thomas Gurney Randall established in 1867 at 6 Elizabeth Terrace, one of three meat trading brothers, who's standing amongst the hierarchy within the meat trade and local community was equal to the Royal households and aristocracy they served with appointments to Queen Victoria, the Prince of Wales, and the Duke of Sutherland. The ultimate high class butchers and respected cattle breeders and judges that served the elite of the residents, schools, and hotels from ten shops in the North West district of London.

Figure 47: The Randall Brothers, 1923.

In 1881 the census record show the Randall business employed fifteen live-in journeymen butchers accommodated between the Englands Lane and Haverstock Hill premises. To this we may add several ancillary staff not living in; such as slaughterhouse workers, stablemen and shop cashiers. A stalwart member of the vestry, Thomas Randall in an association with Hampstead council of nearly fifty years, proposed Sir Henry Harben the first mayor of the borough. He also found time to indulge his social and philanthropic interests. A Master of the butchers company in 1897, Thomas played an active role; on one occasion travelling with a colleague to St Louis on an official visit to the United Masters Butchers Association of America, his character and achievements undeniably impressive.

One of three brothers, Thomas shared the family business responsibilities with Joseph and Henry, which besides an abattoir in Elizabeth Mews, included nine branch shops in Hampstead, Highgate, Finchley, Hornsey and Portman Square and others areas in London. All the shops reputedly to be among the first to be connected by telephone. The Highgate and Hornsey branches were run by Joseph Gurney Randall, who in 1888 became the target of horse rustlers. When two horses he owned were stolen on the evening of July 3rd from a field he rented in Broadlands Road, Highgate. The rustlers George Smith and William Gossage were apprehended, one horse being found in a horse dealers yard in Birmingham the other stabled in Queensland Road, Holloway. It was stated at the trial both animals were of considerable value being high class carriage horses. Of interest, the Highgate branch shop sited at 62 the High Street and currently Café Nero, still retains a wooden canopy extending across the pavement, with the words 'The Old Butchers Shop' as a reminder of its former use. The Randall family originated from Langley, Bucks where their father was a prominent farmer and subsequently moved to northwest London. They were always deeply involved and generous supporters of the guild and on occasions presenting gifts to the Worshipful Company of Butchers.

Figure 48: T G. Randall, Englands Lane, c1906.

The items included a silver tankard embossed with a bull and horse and gilt Warwick vase on ebony plinth. In addition a sculptured head of Joseph stood for many years in the small court room within the Butchers Hall. The trio of brothers when photographed in 1923 had an aggregate age of 241 years.

In the photograph of the Hampstead premises used by Charles O'Hara, we have the unusual, but not rare sight of a shop with two names. For left and right at the top of the facia above the original owners name can be seen Lidstone Ltd., the current occupants at this period, the company history discussed fully in Chapter 13. It is probable the previous business name was essential in order to continue using the Royal warrant; a legitimate if problematical course of action because Charles O'Hara was on the board of Lidstone directors. There is no evidence to which brother the royal warrant was granted albeit Williams undoubted knowledge of a good a butcher's beast was used to great effect as a judge at many of the Smithfield Shows.

Figure 49: C. O'Hara, 17 High Street, Hampstead, c1900.

The O'Hara business of four shops was divided and operated separately by two Brentford born brothers; William the senior by three years controlling shops in Highgate and Leighton Roads, and Charles with shops in Hampstead and Kilburn. The elder William held a lease on premises at 89 Leighton Road from Michaelmas 1882 at a rent of £52 per annum and the freehold of his other butchers shop at 77 Highgate Road.

In the 1880s, the Highgate Road shop and also the family residence provided tied accommodation for various employees. The number of occupants listed for the 1881 census totalled twelve; of these were William, his wife Agnes with children Nellie and William junior. The staff, both butchery and domestic, accounted for eight persons, three butchers and one improver together with a bookkeeper and a children's nurse, one general servant and a butchers-ostler. The last category meaning a stableman, the trade prefix simply meaning the person in this case twenty-one year old Edward Turner was dually employed as delivery butcher and stableman. A similar number of business and domestic staff were employed by Charles for his home and shops, namely a bookkeeper, butchers and nursemaid all living above the premises in High Street, Hampstead. Following the sale of his business, William moved to 73 Bell Street, Henley-on-Thames.

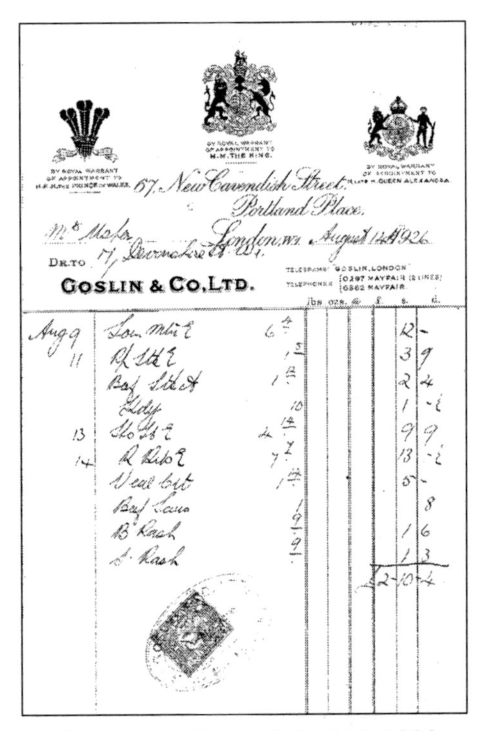
By Royal Warrant of Appointment to H.R.H. The Prince of Wales.

By Royal Warrant of Appointment to H.M. The King.

By Royal Warrant of Appointment to H.M. Queen Alexandra.

67, New Cavendish Street, Portland Place,

London, W1. August 14th 1926

Mrs Usher, 17, Devonshire St. W1.

Dr. to

Goslin & Co. Ltd.

Telegrams "Goslin, London"
Telephones 0297 Mayfair (2 lines) 6562 Mayfair

		lbs	ozs.	@	£	s.	d.
Aug 9	[illegible]	6	4			12	-
11	[illegible]	1	5			3	9

Figure 50: Goslin & Co Ltd. 1926

It is believed the founder of Goslin & Co was butcher James Goslin who traded during the 1830s from number 16 Clare Street. The street wherein John Holles, 2nd Earl of Clare had a market house built in 1647 from which evolved Clare Street Market that by 1853 had become principally a place to purchase butchers meat and vegetables. The market was demolished in 1905 as part of slum clearance under the Kingsway-Aldwych improvement scheme. The business moved to 39 Great Marylebone Street where Thomas Goslin is named as proprietor. Eventually a family partnership was formed between T.C. and H.J.C. Goslin, under the style of Goslin & Co., the partnership was dissolved on the 10th April 1895, at which time included a wholesale business in the Central Meat Market (Smithfield) EC1. After they jettisoned their wholesale interests, the retail side relocated to New Cavendish Street and became incorporated in 1912 as Goslin & Co Ltd, and was still trading from this site in nineteen-forty-nine, in this year holders of the Royal Warrant of long standing and supplying meat and poultry to the Royal household.

A receipted invoice has survived dated 1926, for goods purchased by a Mrs Usher, living in Devonshire Street, St Marylebone, presumably the household cook in the residence of Hebert Arnold Pallant, a dental surgeon and son of Arthur and Henrietta who lived and had a dental practice in Maidenhead. It was in this town Herbert was born and later recorded there aged eighteen an articled pupil in dentistry. On the bill head the insignia of three members of the royal family confirms yet again the customer could shop with confidence at this establishment.

A Royal warrant holder to Queen Victoria and the Prince of Wales was Alfred Stanbridge Ginger trading from three other sites in Edgware Road, Stanmore and High Road Kilburn, still occupying the latter in 1934. He was rather unusual in having the foresight to predict the abnormally fat animals then still in vogue at cattle shows in London and throughout the country during the eighteen-eighties were becoming increasingly unacceptable to the meat buying public. Holding the opinion the proportion of lean to fat on the butchers ideal beast of the future will have to be drastically altered, albeit his views were totally unconnected with health or dietary considerations but based on the resulting

Figure 51: A. S. Ginger 249 Maida Vale, Hampstead, c1893.

large amounts of waste. The way forward he advocated was less slavish devotion to pure strains of cattle and more cross breeds for beef purposes, in short a fleshy type. Historically fat cattle, sheep and pigs were much in demand as source of energy and cooking before the advent of manufactured fats and the provision of gas and electricity utilities.

The firm George Runnicles was established in the 1840s, the shop in Drummond Street, sited near The Duke of Edinburgh public house and as the three feathers in the Royal insignia above the shop implies purveyors of meat to H.R.H The Prince of Wales. On the right of the shop window are two pig carcasses an arms length from garments hanging outside the adjacent cloths shop. The firm had six branches at varying periods including Marchmont Street, the Regents Park shop advertised on the shop blind sited in Osnaburgh Street.

In Camden Town the butchery business of George Grantham, already mentioned, also catered for the needs of royalty. With the prospect of enhanced trade and reputation inevitably some unscrupulous businessmen resorted to fraud and bestowed upon themselves Royal approval without authority, and butcher Tom Crisp trading as Webber & Co in Edgware Road did just that and was fined £15 with £5.5s costs. He was summoned in February 1897 by the Incorporated Association of her Majesty's Warrant Holders Limited, the official guardians of such matters.

Figure 52: G. Runnicles 134 Drummond Street, Euston, 1903

A curiosity of tax duty in the eighteen seventies tradesmen were not taxable for using armorial bearings, or devices in connexion with trade, as on bill heads or trade labels, but were libel for duty if used on paper for general correspondence or otherwise than strictly in connexion with trade. Royal Warrants of appointment are currently administered by the department of Her Majesty's Privy Purse and appointed tradesmen are entitled to display the royal Arms, but not to fly the Royal Standard. The use of the word Royal is strictly controlled by the royal household tradesmen's warrants committee.

MEATHIST MINIATURE

The payments to Yeoman Purveyors of meat to Queen Elizabeth I are recorded in the Royal Housekeeping Book, "Purveyors of Beeves & Mutton to have £1. 13s. 4d. a yeare ald, all riding charges. The Purveyors of Veals 100 shillings a yeare and 20 pence diem Board and Lodging". The first Monarch to allow tradesmen to use the Royal Arms was King William IV. Today the Lord Chamberlain of the Household has overall responsibility for granting the Royal Warrant.

6

BUSINESS AND COMMUNITY

The very existence for the retail butchers trade is the community where population, housing, employment, education, diet and health conspire to determine the number and class of trader attracted to a particular area. That persistent problem of how to 'fit a quart into a pint pot' however, prevents us delving too deeply into the social or environmental issues, other than providing a pot-pourri of one neighbourhood in and around Malden Road in west Kentish Town. Often dismissed as a secondary road, channelling travellers to Hampstead and shoppers to Queens Crescent market, the commercial history of this thoroughfare played an important part in the development of Kentish Town.

The road was built as a series of terraces, beginning in the 1840s on land owned by the Southampton family, and received its present name in 1863 when sub-division names were abolished. These sets of buildings on the east and west side of the road ranged from two houses to thirty, each with its own name and sequence of numbers: Malden, Ponsford, Chambers, St George's, Kingsbridge, and Gipsy Terrace, together with lower and upper Newbury Place and others. The road name derives from their estate in New Malden Surrey, a gift from Charles II to the Fitzroy's, later Earls of Southampton.

When originally proposed, the development plan for Malden Road and surrounding area augured well for an open and pleasant environment in which families had space to grow. However, reality bore scant resemblance to architectural drawings, for in truth on this occasion space was a valuable currency not to be wasted on future inhabitants. To compound an undesirable situation, a succession of bankrupt builders were forced to sell substandard and carcass houses in and around the Malden Road area. Inevitably West Kentish Town became just another copy from the template of other high density populated urban areas, attracting an abundance of trades and professions. In 1891 the census results reveal the population of Kentish Town was three times that of the other five sub districts in St Pancras.

The more that came the more that followed, while nature did her utmost to increase the numbers already there. Isolated pockets of depredation, politically motivated newspaper reporting, disgruntled diarists and a society with a predilection for labelling, served to give the whole area an unfair reputation for uncouthness, almost to the point of being regarded as a separate enclave. Dowdy and down at heel at times during its history it may have been, but considerable wealth was generated from the inhabitants through profitable businesses.

NOAH'S ARK

Of greatest individual environmental impact butchers and urban cow keepers, resembling Noah's Ark trundled two by two, with their stock in trade of cattle, sheep and pigs. There presence in our towns and cities a source of condemnation and controversy. There was a time however, allowing for artistic license, that most enduring pastoral scene of our agricultural heritage, the dairymaid attending a cow at milking time in a rural idyll was an everyday occurrence on farms, smallholdings and cottages. A pleasant and natural representation we can hardly imagine being any other way. In stark contrast to the noise and smell of cow houses, slaughterhouses, stables and general pollution caused by open cess pools, dung heaps, and sewers that surrounded the vanguard of urban dwellers. The rural dairy herds provided an important source of supply from the farmers, who dried off the milch cows when they reached a certain age or milk yield, declined and fattened them up for sale at market.

The urban cow-keepers on the other hand found it more convenient to sell locally, especially the afflicted animals sent to the nearest butcher under the cover of darkness. It was a business opportunity already recognised by some butchers with dairy shops using dual purpose cattle, for example butcher J. Allen, running both operations from the first buildings erected in the Queens Crescent market end. The word milch is old English, and figuratively speaking, meaning a source of profit that gave rise to the term for the milking cow as 'the walking bank'. In consequence those belonging to individual cow keepers and smaller dairies, deprived of pastures spent a short miserable existence tethered to their stalls, or huddled together in confined spaces, a great number kept in yards behind the dairy shops. The average life expectancy of cows in London, for the period 1840-1860 for example, was estimated at five months compared to nearly three years in country areas. The west side of Kentish Town had the biggest concentration of cow houses and no exaggeration to say many residents lived in close proximity to them for the ownership of a single cow was then a common occurrence. By the 1890s they were in the hands of large herd owners like Frederick and Maria Camp, with one hundred and twenty-five cows divided between three locations at Gospel Oak Grove, Ferdinand Place and Shipton Yard, the last named in the rear of 156 Prince of Wales Road. Typical was Evan Benjamin of Modbury Street with fifteen cows in sheds hemmed in on all-sides by domestic dwellings, and in the same predicament Albert and Arthur Kingham of Winchester Yard, Bassett Street twenty-two cows, later to be Golden Dairy Farm.

Attendant to this period was the countrywide adulteration of the milk. One offended observer in 1900 wrote 'I saw the cowman wore an apron, a dirty piece of sacking, greasy with use and an equally greasy cap. The smell of cow muck pervaded his person whilst he drew buckets of water from the iron cow', the latter a euphemism meaning water from a well or water tap to increase his profits'.

The names of traders prosecuted nationwide for selling adulterated food are legion, in particular milk and butter. As in Kentish Town the culprits, often a stones throw away from one another, in Fortess Road dairyman Thomas Draper and cheesemonger Robert Opie together with dairyman Thomas Prenck in Highgate Road. All these shopkeepers were caught on the same day and all fined varying amounts at the magistrate's court. In 1899 Fortess Road provision merchant H C. Rouch was fined £10 pounds with 14s. 6p cost, for selling butter adulterated with 70% foreign fats. Similarly in the same year repeated offender William Evan, Hartland Road, fined £5 with cost for selling milk from which 19% of cream had been extracted, and Emma Griffiths, 324 Kentish

Town Road fined 10s with 12s costs for selling milk with 10% added water. The urban based milk and dairy industry eventually returned to its natural environment, and in 1948 free school milk was introduced, a safe and pure product for future generations.

EARLY ARRIVALS

Of the butchers some of the early arrivals were first timers, others were experienced tradesmen with an eye for business like James Millar, here by 1857 expanding from his shop in Chiswell Street, Finsbury Square, he acquired another in Balls Pond Road (formally Bulls Pond Road) and finally 18 Malden Terrace, also the family home; soon after opening he was granted a slaughterhouse licence. At the time of the 1861 census the family consisted of fifty year old James his wife Mary, two years his senior, and three Lewisham born daughters, seventeen year old twins Isabella and Agnes and Elizabeth aged thirteen. Sharing the accommodation was John Macpherson, an unmarried twenty eight year old butchers-man from Bromley in Kent. For live-in staff, board and lodgings were then part of the wages paid.

As housing expanded, allowing rooms available for rent, the pattern changed and more typical would be locally employed journeyman butcher Harold Duggins from Buckinghamshire, living in Wellesley Road in 1871 with his wife both aged 33 years and six children. Tied accommodation however, while not so prevalent today still remains the carrot on the stick for many a young married butcher. Another early arrival was John Holford in 1855, living in Dukesfield Terrace, a short run of houses in Prince of Wales Road, presumably he preferred and could afford not to live above his shop premises at 30 Ponsford Terrace. His name sake the Holfords of Hampstead feature in Camden's local history and the name occurs in guild records, in the matter of a £150 pound loan made to the Worshipful Company of Butchers by Isaac Holford master in 1704. A common enough name, family connections are thought unlikely. The butchers shop belonging to John was on the corner at the junction with the western side of Queens Crescent re-designated 97 Malden Road when sub-division names were abolished.

After Holfords departure, among other uses, it became home to the Rose and Thistle coffee tavern before being demolished with one other building to make way for Barclays Bank Ltd, (formally the London & Southwest Bank Ltd) that dominated this corner. The bank at first opened at 89 Malden Road on 7th January 1924, almost immediately larger premises were urgently needed, and by nineteen-thirty had relocated. The offices above the new premises were let to dentists and solicitors until the bank eventually closed on 7th December 1973. The building has since been demolished, but it is inconceivable that an alternative use for this fine building could not have been found.

A succession of Ham, Beef & Tongue shops, the forerunners of cooked meats retailers occupied the shop next door, the weather beaten Gilded Ham, the sign of their trade, hung above the shop entrance until the nineteen fifties.

Typical of the pattern of migration for tradesmen arriving in London from the 1840s onwards was Henry Webb from Hampshire. It is unknown if he came of his own volition or as part of a family group direct to Kentish Town; however by 1855 he had opened a butchers shop at 12 Ponsford Terrace where the Webb name survived until c1925. This address was renumbered 61 Malden Road in 1863 when sub-division names were abolished. The family unit consisted of

Henry and wife Rebecca and three children Rebecca, Henry and Charles all of whom were born in St Pancras. The premises were situated on the west side of Malden Road two shops removed north from the Ponsford Arms public house on the corner of Rhyl Street. The public house and original terrace named after Mr Ponsford, the property landlord who also held leases on property in nearby Marsden Street. The details are unknown but from the 1870s Mrs Rebecca Webb is recorded as sole proprietor until at least 1901 and possibly beyond. When at this date recorded as a widowed woman aged seventy years old, employer at home living above the shop. In addition, occupying three rooms were husband and wife Edward a licensed driver and Clara Harding. A family connection is unknown, however it is believed the day to day running of the shop was undertaken by Henry aka Harry Mrs Webb's eldest son. He lived with wife Alice in Grafton Road and one of their children, fifteen year Arthur also worked at the shop.

Other arrivals before the turn of the century included butcher John Williams of Kingsbridge Terrace and Charles Summerlin 28 Newbury Terrace, next door but one to the soap & candle warehouse (shop) of tallow merchant Walton Hassell & Port, and Pork butcher Robert Carter who arrived circa 1855, taking over a defunct wet fish shop at 17 Newbury Place, often spelt as Newberry (later 6 Malden Road),. This shop again reverted to the fish trade under several different owners, selling both wet and fried fish, most notable the Cowen family with other local branches that in 1932 replaced John Weintrop.

By the early 1860s, newly named Malden Road could boast a full complement of trades and professions suited in every respect to furnish the needs of local residents and those yet to arrive. Food and drink sales as always were the driving force, six each of butchers and grocer's shops, three bakers, and two each of dairies, fruiterers, cheesemongers and one greengrocer. With the prospect of additional shopping facilities yet to arrive on the eastern side of Queens Crescent and a street market, the established retailers would have plenty of competition. A future viewed with justifiable concern by the business community in Malden Road, some taking pre-emptive action by renting shops in the newly built Queens Crescent market end. The continual inflow of new residents however abated any worries for by 1874 the number of food retailers had increased to thirty-five and the butchers shops among them had risen to eight.

As to the nature of the community which they served, it is apparent from sample incidents chosen at random from vestry minutes and numerous press reports that our modern equivalent of anti-social behaviour was evident from very early on. The stealing of food was then considered a heinous crime and harshly dealt with, as the following example illustrates. When Edwin Coxan and Henry Oliver, who were both fourteen years old, were found guilty of stealing a chicken belonging to a Mrs William living at 1 Wellesley Road, Kentish Town, both were sentenced in 1866 to fourteen day's hard labour, followed by five years on the reformatory ship Cornwall. Unfortunately for the culprits the magistrate sitting at Marylebone Court recollected having seen them before him on other charges.

In response to an influx of single men, mainly building and railway workers there were a few disorderly houses; disorderly in vestry code covered a multitude of sins. In the case of 17 Dukes Terrace in 1871, then the western part of Rhyl Street, it meant prostitution. The criminal element continued to make themselves a nuisance and in March 1879 lessees of mews in Modbury Street felt obliged to make application to erect gates to prevent thieves and disorderly characters from entry. The vestry agreed, provided they were given a spare key for the lamp lighter. Newspaper reporters wrote dramatically of 'Malden Road Roughs' and other Bill Sykes type characters; violence and drunkenness it was said was endemic and murders averaged one a year. The historical facts are undeniable, but no more so then Somers Town and Camden Town and other areas at differing periods in their development.

One of many troublesome places, being the Lads of the Village beerhouse in Harold Street, Lismore Circus and in particular the Foresters Arms beerhouse in Hampstead Road, a haven for out of hour's drinkers and disorderly conduct. Both establishments visited by police Inspector Harris late on Saturday night in April 1872, where he found drunken men and women.

Figure 53: H. Webb, 61 Malden Road, c1906.

One family tradesman affected by local crime was butcher Robert Roberts and his wife Emma at number 7 Warden Road, arriving in the area from Lambeth in the mid eighteen seventies. Despite having a family of three daughters under the age of five, all of whom were born in Kentish Town, had since their arrival managed to build a steady business. However an incident that began at 7 o'clock on a Monday evening in February 1884 in Roberts shop, Warden Road led to an inquiry of alleged manslaughter against 516Y, Police Constable Mann. On that night Robert, whilst engaged in conversation with Mr Adams licensee of the Carlton Tavern, three men heavily under the influence of drink entered his shop. They began to "pull the meat about and became abusive" when asked to leave a dispute arose and Robert called for assistance from Constable Mann patrolling nearby. A violent scuffle ensued between the officer and one of the men Thomas Badlwin, a horse keeper from Grafton Chambers, Litcham Street, known locally as dung heap street. Mr Badlwin consequently suffered a broken ankle which was attended to by a local Doctor. The injury failed to heal and he subsequently died from complications, the police officer was exonerated. The butcher Robert Roberts eventually formed a partnership with Thomas Stone trading from 102 Queens Crescent.

In 1888 the Y, or Highgate & Upper Holloway, Division of the Metropolitan Police of which Kentish Town was a sub-division, with Superintendent William J. Huntley in charge, numbered 46 inspectors, 4 Sergeants and 147 constables. How many of these police officers were designated for duty in west Kentish Town is unknown. It is recorded a police constable was permanently stationed from 9 a.m. to 1 a.m. near the Mother Shipton public house in Malden Road, and at the junction of Southampton and Circus Roads. The consensus of opinion at the time, including that of the police, concludes the area had generally improved albeit Malden Road was still noisy at night time.

In contrast to ruffians enlightened individuals, with demure sounding names, offered a different future through education to those who cared to apply and could afford the tuition fee. The scholastic academy run by William Temple, augmented by the likes of Miss Jane Hunt's School for Young Ladies, and Benjamin Desmarest teacher of French, were first on hand to feed the mind, and if inclined artistic ambitions were satisfied at the Stage College of Dramatic Tuition founded in 1866 by Mr Henry Leslie, 36 Queens Crescent, Kentish Town. A more formal education arrived with St Andrews National Schools, providentially next to Carl Gottlieh the bootmaker, whilst nearby the William Ellis School, c1864, was available in Allcroft Road. The opening of board schools in Haverstock Hill, Holmes Road, Carlton Road and Rhyl Street, with a combined total of 2,652 pupil places, fulfilled a yawning gap in local education and gave the area a real sense of community. Whilst free and unfettered access to reading material in public libraries arrived belatedly in 1904, the first in Malden Road opened in 1946 over forty-years later; the site chosen a disused boot makers, for many years prior Spencer & Co the drapers. For the habitual drinker the inevitable public houses, five in total, were strategically sited at convenient intervals along the road, offering a bill of fare including milk, tea or coffee and a choice of beef, mutton or sausage sandwiches.

Figure 54: Elvidge Bros, 37 Malden Road, c1905.

As it transpired, from the beginning pawnbroker James Mellish at 50/52 Malden Road was of more practical assistance. The service, by familiarity known as 'uncles', continued an unbroken line of local need from this address for over one hundred years, the last incumbent being jeweller and pawnbroker John Long Ltd. The services of money lenders and public house loan clubs provided an additional source which maybe said has a chequered history; for the most part illegally enforcing heavy fines on the borrower or taking securities on the non payment of instalments. A state of affairs that in 1840 compelled the government to pass the Loan Society Act that remained on the statute books until repealed in 1989. A significant number of loan clubs and societies by the eighteen sixties were based in public houses: The Crown, Bull and Gate, Tally Ho Tavern, Black Horse, Gospel, Grafton Arms and Gypsy Queen and all in Kentish Town and all hosts to the Labour Loan Society.

The name Sir Robert Peel (1788-1850), chosen for one public house on the east side at the junction with Queens Crescent, has particular pertinence. Although best remembered for establishing the first police force, the 'Peelers', and penal reforms whilst Home Secretary; as Prime Minister from 1841 to 1846 he was equally influential in repealing the Corn Laws and matters affecting the meat trade. It was said he was nicknamed the refrigerator man, for his support of the ice trade although his political opponents insisted the name was actually a reference to his coldness towards parliamentary colleagues. Sir Robert was also instrumental, with the Duke of Wellington, for the foundation in 1838 of the Royal Agricultural Society of England; motto Practice with Science.

When Kentish Town was larger in area than it is today the edifices to religion were the main meeting places for folk to gather; that until urbanisation served the community when a plethora of mission halls staked their claim for future expansion. From which grew in Malden Road and immediate vicinity the churches St Andrews, St Silas the Martyr, St Martins, St Dominic's and others.

A civic duty or belief that cleanliness is next to godliness, but oft times attributed to the daily bathing ritual of the Duke of Wellington. From the late 1880s a surge of public lavatories, laundry, bathing and swimming facilities were built; these sanitary amenities, hitherto woefully lacking in Kentish Town, although spending a penny had its critics and later G. B. Shaw battled for a free public lavatory in Camden High Street. A vestryman (1897-1900), borough councillor (1900-1903) in 1946 he was made Freeman of St Pancras.

In this regard the changing social situation bought opportunities for other industries and services than the meat trade. The soap makers historically allied via tallow melters seized the favourable conditions and increased their products in variety and number.

Returning to the meat trade fraternity a new generation of butchers Robert and Arthur trading as the Elvidge Brothers, were installed at 37 Malden Road shop c1885, exploiting imports of frozen meat. The premises were adjacent to Steeds grocery shop on the western side, on the corner of Marsden Street facing the Newbury Arms public house, recently replaced by an apartment block. Entrance to the rear of the shop and slaughtering area were through double wooden gates immediately to the right of the window that faced onto Marsden Street. The accommodation above occupied by Robert the eldest born in 1866 with wife Mary, son Allan both assisting in the business and school age daughters Winifred and Muriel.

The two brothers also had branches in Harmood Street, Kentish Town Road and Caledonian Road, all with licensed slaughtering facilities. For reasons unknown, prior to the First World War all three branches in Kentish Town had ceased trading, an oilman and grocer had taken over two of the shops and the Malden Road branch by 1915 had become Mrs Florence Walkers confectioners. The Malden Road shop ending its remaining years, from 1938 until demolished, as Radford's toy shop; Elizabeth Radford worked in the shop assisted occasionally by husband Ernest, who had employment elsewhere. At the time of this photograph the butchers shops and market stalls nearby in Queens Crescent had all but taken the lion's share of trade from Malden Road. Yet there was still a reasonable living to be earned, particularly if the meat was of good quality as market traders had a reputation, not always justified, for inferior imported meat and sleight of hand.

Much of the poverty and crime was attributed to low incomes and the intolerable living conditions that existed around Malden Road and other parts of west Kentish Town. A world of slop buckets, mice occasional rats, bugs, other infestations and their avowed enemies, traps, poison, flit, fly paper, sulphur bombs, carbolic soap and disinfectant. In this regard we may touch upon the work of noted Victorian social reformer Charles Booth (1840-1916), and his team of investigators. The son of a well to do Liverpool family made prosperous from involvement in shipping; he published a series of district street maps and data in 1899 covering London. The maps coloured according to the levels of poverty the lowest class denoted in black. The gold standard (yellow) upper middle classes and wealthy, a rare find in any part of Kentish Town. His sombre colouring of some west Kentish Town streets only confirming to the inhabitants living there what they already knew. The Rev G. Napier Wittingham, curate of St Silas Mission Church, using the phrase 'neglected and overlooked Kentish Town' during his appeal for donations to provided a day out in the country or seaside for children in need.

A personal if slightly disenchanted view is given by Alfred Grosch, in his autobiographical book St Pancras Pavements, in which a few of the 280 pages are devoted to the surrounding environment of 10 Malden Road, where his family resided and he freely admits made a good living as corn and seed dealers. His comments express an obvious distaste for the area in which he spent his childhood years where, 'the people were brutal and pugnacious' and 'a dozen slum streets were within a stones throw from our home', a world far removed from the confines of his own polite society. His father William Grosch aged 44 is recorded there in 1901 as corn dealer and his son Alfred as oil shop assistant. The shop was previously owned by Thomas Tasker & Co, one of four branches located at Dartmouth Park Road, York Road and Camden Road. The firm commenced business in 1864 dealing in corn, hay and straw until June 1891, when the business was declared bankrupt. The Newbury Mews, with a side entrance to the Grosch home was used for a variety of trades and clandestine activities, including illegal slaughtering and gambling was home to costermongers, market traders with their families and horses, carts, stalls, barrows and stocks of fish, meat, poultry, game, fruit, flowers and vegetables. To the outside local world at large the activities within remained a mystery, a place where sensible adults avoided after dark, and streetwise children learned never to enter unless they were open ended. Time has since rung the changes, and today they are smart, sort after places, transformed by property developers and residents into an oasis of tranquillity with colourful flower filled tubs, window boxes and motionless painted cart wheels that once clattered upon the cobble stones.

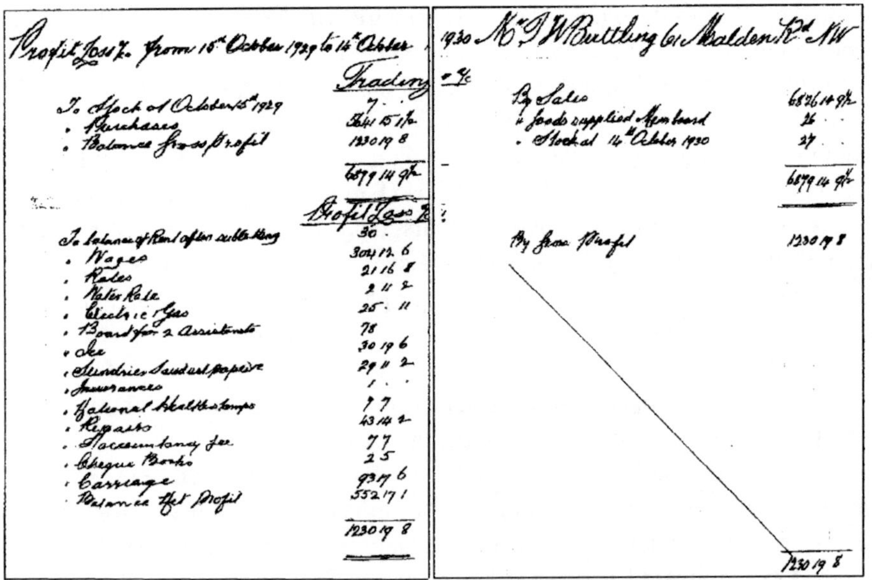
Profit & Loss a/c from 15th October 1929 to 14th October 1930 Mr T W Buttling 61 Malden Rd NW

Trading a/c

To Stock at October 15th 1929	7 . .	By Sales	6826 14 9½
" Purchases	5641 15 1½	" Goods supplied Mem board	26 . .
" Balance Gross Profit	1230 19 8	" Stock at 14th October 1930	27 . .
	6879 14 9½		6879 14 9½

Profit & Loss a/c

To balance of Rent after subletting	30 .	By Gross Profit	1230 19 8
" Wages	304 12 6		
" Rates	21 16 8		
" Water Rate	2 11 2		
" Electric & Gas	25 . 11		
" Board for 2 Assistants	78		
" Ice	30 19 6		
" Sundries Sawdust Paper &c	29 11 2		
" Insurances	1 . .		
" National Health Stamps	7 7		
" Repairs	43 14 2		
" Accountancy fee	7 7		
" Cheque Books	2 5		
" Carriage	93 17 6		
" Balance Net Profit	552 17 1		
	1230 19 8		1230 19 8

Figure 55: Profit & Loss a/c, T. W. Buttling, 61 Malden Road, 1929/30.

On balance the economic picture of Malden Road area, and for that matter Kentish Town, showed average earnings above the line of poverty and in some instances nudging the classification of lower middle class; those in this group managed a roast on Sundays and some meat or fish meals during the week. For in truth from the 1870s, imports of wheat for bread, meat and dairy products had steadily increased and gradually became affordable to those on the lowest income. That is not to say people did not endure personal hardships when the national economic temperature plunged violently, but this was not peculiar to Kentish Town. The fortune of any area wax and wane, but generally speaking by 1900 food was plentiful and cheap. The higher earners availed themselves of the more varied diet available and choicer cuts of meat while the working classes relied on appetising meat and basic vegetables. For this was still the era of large families with anything approaching twelve children, but with crucially a significantly higher survival rate. With physically demanding manual work for both men and women, it was a time of the 'Sunday Oven Busters', great lumps of fatty meat accompanied by mounds of vegetables, washed down with a mug of hot green water. The name derived from the watery liquid strained from cooked cabbage and greens.

Yet the children still raked over the piles of rotting produce in Queens Crescent market, a practice that was still evident in the 1950s, then looking for 'Specks' a euphemism for badly bruised and heavily speckled apples. For that is their nature, always hungry, always inquisitive and always the hunter, inherent in the human male. Any attempt at comparative analysis of living standards between west and east Kentish Town in a few pages would be impractical. It does however raise the question whether the evidence was ever powerful enough for a general stigmatisation of the whole area.

The firm of Thomas Wooster Buttling has the distinction of being among the longest serving retail butchers shops in Queens Crescent market. A family history within the trade that began four generations ago in the 1890s, when founder William Buttling in a joint venture with his cousin opened for business in Stepney, East London. The business association with Kentish Town did not begin in the market however, but at 61 Malden Road in 1928, a small yet successful shop with a strong customer loyalty base providing a springboard for expansion. We are fortunate a profit and loss account has survived from the period that among other items shown are 'Board for Assistants' and the considerable cost of 'Ice'. In 1934 the opportunity arose to lease and eventually purchase the freehold of a shop previously owned by James Rayner Ltd in the nearby busy market street. It was from here, sited on the corner of Basset Street that began a tenure lasting fifty-one years and the induction of another generation into the family business.

Figure 56: Bennet Wright, 1971.

The new entrant and third generation Tom Buttling, now in his seventies recalls, 'I worked in my fathers business from 1946 at Queens Crescent, that I remember had lots of decorative tiles, and after national service I went to manage our branch at Harlesden'. By the 1960s, following the ill health of his father, Tom assumed the mantle of responsibility for buying and general administration, which at this period besides Kentish Town included branches at Camden Town,

Holloway and Edgware. The family were obviously good employers, as the longevity and dedication of loyal staff member Bennet Wright with forty-one years service, bears testament. Joining the firm straight from school and working mainly at their Malden Road branch, he became the epitome of the cheerful and helpful butcher. In 1971, reflecting on the imminent closure of the shop due to redevelopment and still with his infectious smile he spoke fondly about his early years. "I can go back to the days when trams used to come up the road in 1926; and not so long ago it used to be so busy you couldn't cross the road". Today pedestrian and road traffic has increased tenfold but many of the shops and shoppers have long since disappeared. It is fitting Thomas Buttling should be the last occupant of this butchers shop, which had traded as such for over a century. There coincided, with the redevelopment of inner cities during the seventies, the beginning of a widespread decline in independent and company retail butchers and by the nineteen-eighties Tom Buttling had relinquished his business interests in the area. However far from ending the history of this family business a new chapter continues in Dorset, where son Paul a fourth generation of Buttling retail butchers continues to this day.

VICTORIAN GENTLEMAN

Throughout its history Malden Road, with the exception of a London Cooperative Society grocery shop of short duration on the corner of Shipton Place in the 1920s, has never managed to entice the big names of food retailing. Instead remaining exclusively the province of the small retailer, men like Ernest Robert Vincett, a butcher of note trading under his family name at 51, situated on the west side in the 1950s between Thomas & Reed carpet sales and repairs, and A. B. Hemmings Ltd., the gold medal bakers. This well known bakers and confectioner was established in 1906 by Mr A. B. Hemmings as a single shop at 601 King's Road, Chelsea and became a major player in the industry; retail outlets totalled one hundred and thirteen by 1947 with seven main bakery units. Recorded at the Vincett address in 1969 are P & G Butchers, if correct their tenure would have been of short duration. The shop had originally been a cheesemongers until the late 1890's, before changing trades to become a shirt and collar dressers. In common with other shops, suffering various internal alterations to suit the occupants, it was altered yet again in c1922 to accommodate William Keen the first retail butcher to trade there. Functional rather than decorative, the frontage consisted of the now familiar metal roller incorporating a single removable door. The window, set to one side with a marble base, and inside a refrigerator positioned at the rear and the usual trade furniture all on one level. Ernest purchased the shop in c1929 and stayed for over 30 years, surviving the restrictions under Meat Control during the Second World War only to succumb, along with many more buildings, to the bulldozers during the partial redevelopment of this area.

Born in 1881 Ernest was tall distinguished and rather stern looking, the epitome of a late Victorian gentleman. A disciplinarian by nature and upbringing, he was ill at ease with what maybe called the forthright manner and bawdy humour of some of his women customers. His lineage in the trade extended back to Samuel Vincett 1644 -1698, his son also named Samuel and grandson Peter were all butchers in Cranbrook, Kent. A later generation headed by Anthony Vincett, born in 1720, owned a butchers shop in Northiam, East Sussex. On his death in 1797 he left the not inconsiderable sum of two thousand pounds. His eldest son Anthony continued the business acquiring a 15 acre estate and farm, later sold during the19th century. His accumulated wealth was eventually distributed among the family, a Thomas Vincett receiving a seventh share. He moved to nearby Brede, where he bought a butchers shop and farmland, adding another shop at Bexhill. As is the way, the business was passed on to Stephen, Ernest's grandfather, who sold the shop in 1840 and took himself off to Hythe where he opened a large open double fronted shop and slaughterhouse, using Romney Marsh as grazing pastures for his animals. By the late 1870's, the nomadic Vincett family had moved again, this time to St Leonards Hastings. It was here young Ernest was born and on leaving school, together with his father Frederick, made their way to London and settled in Stoke Newington in 1895. In that same year, aged 14, he entered the meat trade proper as a butcher's assistant, gradually learning his craft in a variety of jobs in north and south London. His natural ability, the result of generations of Vincett butchers, and his own endeavours, ensured a bright future.

Before the new century had entered it's first year Ernest was appointed manager of Bentleys butchers in Brondesbury, NW6, A considerable achievement given his age and he made his first purchase on Smithfield Market in 1900, when he was just nineteen years old. As he recalled in an interview some seventy years later, 'The purchase consisted of ox liver at 3d. per pound and 3½ d. for 'plate' lambs'. The word plate, a prefix used to denote meat imported from the River Plate area of Argentina. From this point on little is known of Mr Vincett's career until c1912 when he took over his first shop at 116a Fortess Road, NW5, and by 1920 added another in Hammers Lane, Edgware.

Outside the practicalities of his retail shops, Ernest was involved with meat trade organisations becoming chairman of division two of the Retail Meat Traders Association and represented them as a member of the bowls team. A family man, his twin sons Phillip and Jack both assisted their father in the daily running of the business, Jack using his motorcycle and sidecar to deliver customer's meat orders. Older readers may recall the vehicle was regularly parked outside the Malden Road branch. The late Jack Vincett, like his father before him, held office in the London association as secretary although his brother Phillip retired long ago. The grand old man himself could still be seen buying on Smithfield Market a few years before he died in 1980, aged ninety-nine. Regrettably after more than three hundred years service, the Vincett's family name has disappeared from the trade scene.

During the 1960-1970s in a phase of redevelopment which went too far, opinions are divided on this point, the Vincett shop along with other businesses and domestic dwellings, including the house where the writer of these lines was born were swept away.

Figure 57: Ernest Vincett, Smithfield Market, 1968.

The corrugated steel fencing that hid the demolition revealing rather poignantly a summer cottage that once belonged to Joseph Stinton. This secret hideaway, known only to a very few, was accessed via a short passageway by the side of his tiny chandlers turned grocers shop at 45 Rhyl Street, the frontage of which was no more than five feet.

The wholesale destruction of Malden Road was averted by public disquiet and the western side, beyond the junction with Queens Crescent, was halted. In this section of the road Walter White, at 157, carried on the tradition of independent retail butcher which had begun in the early 1860s, five proprietors preceding him at this location. In the interior photograph of the shop Walter is seen preparing the joints of meat which in this instance are back and top ribs of beef for roasting. Assisted in the family business by his sons, he was also owner of another butchers shop in nearby Southampton Road in the nineteen fifties and sixties.

SERVICE BEFORE SELF

Beyond the purely commercial aspects of shop keeping lay the social implications. Those who derived their living from the inhabitants of St Pancras often engaged themselves in charitable works. These ranged from individual acts of kindness, to voluntary service on the various institutions and organisations dedicated to the relief of hardship and suffering, benefiting the whole community in most cases beyond the gaze of publicity. There are countless stories of such good intent throughout history ranging from the commonplace known as cow charities where 'the poorest and godliest' had a cow loaned to them for a year at a nominal sum. To the unusual where a Buckinghamshire butcher and cattle farmer, by the terms of his will, left £6 for the purchase of a bull to be baited each year; the proceeds of the event to be distributed amongst the poor children in the parish. An unspeakable spectacle, bull baiting was commonly indulged in the yards and fields of Marylebone, Clerkenwell, Westminster and Soho and other districts in London.

Far from exceptional Kentish Town butchers Liveryman Edwin Cox, resident at 11 Rochester Terrace at the time of his death on 1st July 1884; a businessman of considerable wealth by the terms of his will, Edwin had instructed the executers of his personal estate which amounted to upwards of £35,000 to be bequeath £500 each to the Butchers' Charitable Institution, now (Butchers and Drovers Charitable Institution) in aid of the building fund; the Royal Free Hospital, Earlswood Idiot Asylum, Deaf and Dumb Asylum, Old Kent Road, the Blind School, St George's in the East, the City of London Trust Society, the National Benevolent Institution, the Hospital for Consumption and Diseases of the Chest at Brompton, the Cancer Hospital, Middlesex Hospital, Westminster Hospital, Charing Cross Hospital, and St Mary's Hospital, and legacies for his relatives and provisions for his housekeeper.

Occasionally the intervention of individual philanthropists at local level could have a striking impact, as in the case of Kentish Town butcher George Frederick Kimber, who had accrued considerable wealth from his business and other properties he owned in Bayham Street, Camden Town and elsewhere. The following story speaks for itself, as George was a member of the Rotary Club of St Pancras; motto 'Service before Self', a regular feature being the monthly meeting and dinner held at the Royal Northern Hotel, Kings Cross.

On one particular occasion, a Dr Beaumont addressed the President and fellow Rotarians present on the subject of the recent illness of King George V during the winter of 1928, from which he had now fully recovered. It was said a major contributory factor was the use of Electro-Therapeutic treatment. This revelation obviously inspired an idea in Mr Kimber's mind, and he subsequently made the following announcement at a club meeting, here I quote George himself: 'I've been thinking things over lately Mr President, about the King and his miraculous recovery under the sun-ray treatment. You know, I've had a fairly busy life and had to keep my nose to the grindstone, and not much time to attend to much else, I've not done so badly, so I would like to take a hand in something outside my business now. I've been thinking it would be a grand thing if the poorest child in St Pancras, when needing it, could get just as good treatment as King George. We ought to have available here in St Pancras that sunray treatment that even the neediest could have. We should start an establishment for that kind of thing, get others to help of course, but I'm willing to start a fund with £10,000.' George personally donated that amount, which in today's values would be £282,500, and organised the whole project.

His sun-ray idea became a reality, and in March 1930 the Institute of Ray Therapy and Electric Therapy at 152 Camden Road NW1, was officially opened by the Lord Mayor of London, accompanied by other dignitaries. By coincidence or design the adjacent premises already housed the Hospital Savings Association. The Princess Royal, the Patroness of the Institute, was informed on a visit in 1933 that in a six month period 49,809 treatments had been carried out and 982 new patients had been enrolled. In 1938 the Princess Royal made a return visit to open an extension to the building to provide increased facilities for the electro-treatment such was proved the demand for the service. The institute survived until c1955, re-designated the Medical Rehabilitation Centre specializing in industrial injuries, altered again to the Camden Clinic it continued serving the community, which after all was George Kimber's basic intention.

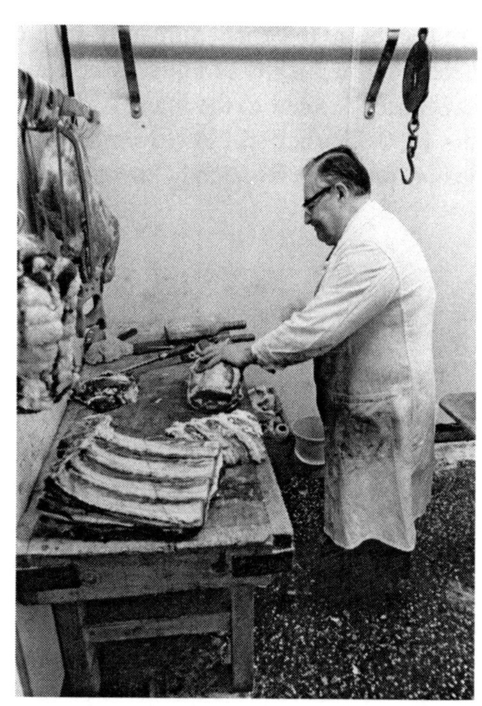

Figure 58: W. White Butchers, 159 Malden Road, 1974.

Reminiscing some years later in nineteen forty, a past secretary of the club, remarked, 'George Kimber startled us all when it was announced he would start the Ray Therapy Clinic, and find the necessary finances. What wonderful work the clinic has done only he can say.' The Rotary Club of St Pancras, one of a network worldwide, was founded on March 25th 1924, at the instigation of the late Mr Stanley Leverton, of the long established and respected family firm of Leverton & Sons, funeral directors.

Not all gifts were of the magnitude donated by the benevolent George Kimber. For many recipients a simple act of charity as performed by Thomas Butting and many other butchers could be of immense help; as defined by Thomas giving a large turkey each Christmas for the benefit of residents at the Alms Houses adjacent to the Priory Church Southampton Road.

In Camden Town from the nearby Convent of the Helpless of the Holy, cherub faced Nuns in black habits beguiling their tenacity went about their daily rounds of shops, actively seeking out donations. They too received gifts of meat and poultry at Christmastides from Harrison butchers in the high street, and no doubt other like minded retailers in this road. In what maybe described as the best of both worlds, ex-butcher the Rev. W. Cuff, of Shoreditch Tabernacle, celebrated Christmas with the distribution of free beef amongst the poor of the district; the meat purchased from funds he collected and personally prepared each Christmas Eve using the tools of his previous occupation. A charitable task he undertook each year from 1875 and one he continued into old age 40 years later.

Unlike the charity gesture of master butcher Frederick W. Griffiths in paying and arranging on Christmas day morning 1889, for a Salvation Army band to play for the benefit of local residents and it was said his own gratification outside his home and shop at 54 Red Lion Street, Holborn. As events transpired not all the residents were of the same mind and following complaints of noise, band member William Custree was arrested for obstruction by Police Constable Williams, 428E, and fined 20s or 14 days imprisonment in default.

The desire to benefit humanity, far from diminishing is on the increase, hence the continuing need for The Society of Philanthropists. With a monthly journal, 'The Philanthropist' already in existence, devoted to the interest of hospitals, charities, and other like institutions, published in the 1880s. Earlier still in November 1843 a benefit was held in aid of St Pancras Philanthropic Institution at the Rose and Crown, Tottenham Court Road.

MEATHIST MINIATURE

The sale and availability of Butchers beef, pork and lamb mincemeat was sporadic before the nineteen – twenties Prior to this the process of pounding and chopping was a tiresome kitchen activity. With the introduction of trade and domestic hand operated machines similar to those manufactured by Spong & Co of High Holborn, London, established in 1856, the process became less laborious and infinitely quicker.

7

SPOILT FOR CHOICE

The one aspect above all others enshrined in the history of retailing is street market shopping, and the truism that shopping revolutions may change the circumstances, but the physical act of shopping in the open air without structural protection from the elements rekindles our inherent need to forage and hunt for food. A principle of human survival that has existed since man first inhabited the earth. In every city, town and village this ancient method of buying and selling was conducted and in many instances responsible for the growth and wealth of a particular area.

The origins of markets are indelibly linked with religious fairs and both were a major development of commercial activity in the middle ages. They differ somewhat in respect that fairs tended to focus on one commodity and were held annually, for example the great Bartholomew Cloth Fair at Smithfield. It was often said "Every fair is a market, but not every market is a fair". Attendant at these fairs were side shows and amusements which gradually took over, becoming principally places of entertainment on public holidays like Hampstead Heath fair from 1712, 'So honestly wholesome, and amusingly vulgar' in the opinion of one commentator' ; official bank holidays did not begin until 1871.

How local markets came into existence as distinct from fairs is uncertain, but being of great antiquity and the majority in later years held on a weekly basis there is little doubt. A wide variety of goods and produce were bought in by people from surrounding areas, who came to sell and buy livestock, cereals, clothing, and leather goods including meat. Of special interest, markets and fairs were once designated as the only legal place for sales to be made as private deals outside, these locations to be legal and binding required two or three witnesses and an official observer. Disputes and complaints between traders and buyers were settled by the 'Court of Piepowder or pied poudres literally meaning dusty feet, as those people with grievances were given justice on the spot; the inference meaning they still had dust from the fairground on their feet.

The granting of charters and franchises for markets and fairs were the prerogative of the crown, ensuring revenues were received and law and order maintained; this system of control remained virtually unchanged from Anglo-Saxon times, until the passing of the Local Government Act of 1856, which conferred on civic authorities the power to establish local markets. It follows that the overwhelming proportion of London markets were not constituted under Royal Charter. Instead where a rapid increase of the population occurred, as in London, traders colonised a specific road or street creating informal markets. This happened wherever there was trade to be had along the main high street or near the departure and arrival points, where horse and coaches travelled to and from London. The right to sell goods in the Metropolis from stands/stalls merely allowing freedom of behaviour whereby from habit costermongers or street dealers had been standing to sell their goods as opposed to the travelling hawker and peddler. They were generally free from police interference unless causing an obstruction or complaint from adjacent shopkeeper or other inhabitant. The law that permitted butchers and other provisions mongers to hire stalls in the flesh and other markets, recognised by custom and usage gave no such permission as to street trading.

In Queens Crescent costermongers and other street traders began to congregate and supplement the ever growing number of shops in Kentish Town. The road straddles west to east of Malden Road, the western part from Prince of Wales Road was built in the 1840s as a series of terraces and named Queen's Road. Building on the eastern side began later in the early 1860s, a section of which was called Cavendish Street; eventually the entire road together with Queens Road was re-titled Queens Crescent. The eastern part of the road 'the market end' soon became the dominant local shopping area in addition to attracting shoppers from outlying districts. This venue together with Kentish Town high street, Malden and Fortess Roads resident shoppers soon found themselves in the enviable position of being spoilt for choice.

The first butcher's shop recorded in Queens Crescent, by virtue of the sequence of building, was not in the market end. This was opened by Henry James Honiball in c1854 at 7 Queens Road, re-designated in the 1870's, as 14 Queens Crescent. The shop was situated at the corner of Rhyl Street, near the Queen's Arms, the public house then number one in a short terrace. It is probable that Henry, born in Clerkenwell, and his wife Martha from Dorset were the first occupants of this newly built terraced house, their children Amelia, Louisa, Ellen and Henry junior were all born to them while they were resident here and by 1871 nineteen year old butchers assistant Henry R. Peabody is in residence. On 4th Junes 1879 Henry was elected vestryman for St Pancras ward 1, and was living at 6 Bellina Villes in Fortess Road until February that year; his father William had premises at 3 Copenhagen Street N1. However by 1883 Honiball the younger had vacated the shop in Queen Crescent and relocated to 790 Holloway Road. We can only surmise that competition from new-comers in the market end of the road influenced his decision to leave. A possibility because thereafter, aside from an off-licence, dairy and a few other trades, this part of Queens Crescent has remained primarily residential, hence the colloquial name 'the private end'.

This is not to state all manner of workers did not reside here and many with local employment. The large houses that fronted the road deemed suitable by the owners from a financial aspect for multiple occupancy. For example the ten roomed house at number 121 on the north side, wherein lived employed butcher Thomas Wormald from Leeds, wife Emily and children Dorothy and Leslie, all in three rooms. In two rooms John and Ada Summers together with John junior, his father a railway post messenger. Completing the number of occupants; mother and son, Mary and Charles Brent, with three rooms his middle aged mother an office cleaner, he working as postman, and lastly Mary Preston and her factory–hand daughter Maud in two rooms; her mother worked as a charwoman and making the unusual, but not unbelievable declaration that she was born during the sea voyage from Ireland. The occupations of near neighbours included solicitors, railways clerks, a commercial traveller, headmistress, piano maker and a choral vicar to the Dean of St Pauls.

Of the original four retail butchers in the market end, George Lacey, J. Allen and T. Nickels; one Mr A. Farey, a pork butcher and tripe dresser of 72 Queens Crescent, announced his arrival with an advertisement in the Camden and Kentish Town Gazette dated 29th September 1866. That he "begs to inform the inhabitants of Kentish Town and it's vicinity that he has commenced business at the above address and hopes to receive their patronage and kind recommendations", his occupancy lasting eight years superseded by another pork butcher, J. Jackson. A pattern repeated over the coming years, as a succession of butchers and a diversity of other retail trades served their time and left for one reason or another.

Yet indicative of past shopping habits and importance of meat, there were never less than five butchers' shops in this road able to fulfil customer requirements; although surprisingly the opposite is true of the fishmonger shop where apart from fish stalls only a sprinkling have set-shop in the market. The number of butchers shops remained constant from 1900 until the outbreak of the Second World War, from which only a representative sample can be mentioned here, some by name only: Thorn, Hobbs, Andrews, Woolton and Bonham. Many like Thomas Charles Hale, tripe dresser, butcher and descendant of the ubiquitous family of local butchers and grocers, who ostensibly retired and passed the business onto the next generation and subsequently became the unseen hand guiding the business. In this particular instance his two sons Thomas the eldest and younger brother Charles, who eventually traded on his own account at 160 Kentish Town Road. Home for this particular branch of the Hale family in 1890 was above the shop premises at 76 Queens Crescent, their four children and staff of varying ages from sixteen to thirty-four with occupations within and independent of the business. In less than ten years from this date Thomas senior at the relatively young age of fifty years old had relocated with his wife Elizabeth to 'The Lindons', their retirement house situated in the village of Shenfield, Essex.

Of significance, adjacent was the shop of master butcher Frederick King and head of a household totalling thirteen people, of these excluding his wife Eleanor were seven children and four domestic staff. With so many young people living and working in such close proximity marriage between the two families was inevitable.

THE MEAT PALACE
155, QUEEN'S CRESCENT, N.W.5. (Corner of Allcroft Road).
R. R. WARREN,
Cash Butcher at Lowest Possible Prices.
And at 26 Grand Parade, Harringay.

Figure 59: R R. Warren, 155 Queens Crescent, 1919.

An intimate partnership between the local meat trading fraternities that extended to included the families of butchers Rayner, Piper and Wyatt. The latter named trading in other areas of Camden, but all these families connected by marriage. The Hale family in this market road are last recorded c1954, as a partnership between Hale & Piper his partner, also a tripe dresser as Piper & Sons had a substantial business in Cowcross Street, Smithfield. The Queens Crescent shop, following a brief spell as May & Shepherd offal salesmen until the arrival of new proprietor George Bayliss a traditional butcher, had come full circle and returned to its roots until finally closing as such in the nineteen eighties.

On the subject of marriage between meat trade families it was the custom of fellow tradesmen to provide musical accompaniment as part of the celebrations. The word music loosely describing a band organised by butchers who created a cacophony of noise by bashing marrow bones and cleavers together. One such occasion attracting the attention of the police when in October 1874, master butcher George Bennett of Leighton Road, marched with his band along Kentish Town road to the residence of the bride's father. The spectacle it was said attracted 500 people and also the police who took Mr Bennett into custody and charged him with disturbing the peace for which he was fined £10 pounds with his assurance to keep the peace for six months. This ancient ceremony until recently was commonly practiced within the close knit community of Smithfield Market, albeit it is rarely witnessed elsewhere.

Figure 60: Queens Crescent, Street Market, c1909.

Others like Alfred Keen trading in the 1880s from number 56a in this market road only a brief sketch is possible. In his late twenties Alfred from Chesham in Buckinghamshire lived above the shop with wife Kate and baby daughter Gertrude. His two butcher brothers, James nineteen and William aged fourteen together with two equally young employee butchers Maulkin and Murray also lived on the premises. The employment of young servant girls was then commonplace and eighteen year old general domestic Martha Martin from East Middlesex completed the number of staff.

One of the more unusual titles encountered for a butchers shop then 'The Meat Palace' used by proprietor Robert Warren in 1919, was arguably an inference to the popularity of the picture palaces, and their obsession with pseudo Roman and Greek architecture and names.

This butcher shared a pecuniary interest in entertaining the public with Saturday night barkers at the auctions, less enduring than cinemas the firm closed after four years. Of equal brief duration, at a later date, was the appearance of such titles as the 'Meat Boutique' designed to convey an upbeat trendy butchers shop, comparable to the then fashionable Carnaby Street and the Beatles pop group era. The first of these so called specialised (luxury) meat shops believed to be that opened in South Kensington by Martin Bayliss c1970, the son of butcher George Bayliss and did initially have the desired effect of attracting considerable local and trade publicity. Nearly twenty years on the title was resurrected for use by butchers at Unit 7, 105 Highgate Road. Other butchers shops have since entered the fray with repetitively indistinctive sounding names: Meat Inn, Meatland, Meat Master, My Meat, Mr Meats and Economy Meats The last named, a second shop opened in Queens Crescent by George Bayliss senior, a former manager of Ritchies butchers who first opened at number seventy six in 1954 trading under his own name.

To avoid confusion, some among the multitude of independent butchers as did the multiples, found it made sense to incorporate their own particular district within the titles, for example Drakes (Kentish Town) Ltd., also a Crescent trader. However, then as now the national preference for independent butchers is still the proprietor's family name above the shop; in this context and rather unusual Hammett & Grandson Ltd; one of the rarer examples of family titles.

In a world now littered with sound-bites and unintelligible slogans the titles hardly raise an eye brow. In the 12th century butchers had no need for business names, being called John the butcher or Jack the slaughterer and so on; eventually these evolved to become the persons surname associated with their trade or calling. These names unintentionally gave rise to amusement and became a bane to their owners, as a selection from the masters guild list covering 1611 to 1916 demonstrates: Cleaver, Mouse, Pigeon, Mallard, Goater, Slaughter, Silverside, Giblet, Game and Hart. Others such as Gamble, a sheep hook, Avery, a weighing scale, Neate as in neats foot oil, have more relevance to those within the trade. A frivolous subject, the unfortunately named Thomas Hoggflesh, the named person discovered quite by chance, as being chosen with two other St Pancras inhabitants in 1643 to assist Dr Denison the parish vicar in collecting the tithes (taxes) and rents.

The North Western Meat Co, here in 1903, dealing exclusively in frozen and chilled meat and managed by Essex born Ernest Bedford living above the shop, there is no evidence this company was ever incorporated, consequently no official records exist. However, we are fortunate in that an archive photograph (not shown), taken of the road c1904 shows in the extreme left hand foreground corner a side blind attached to the shop at 58 Queens Crescent, on which is printed the company title. The pork butcher Rayners Ltd, who seem to have adopted a variation of the favourite Victorian parlour game musical chairs but using shops instead, were here just before Harry Denby Ltd. Based in Camberwell and known locally known as Denbys, this company departed in 1958 and were dissolved in 1973. After a short spell as Central Stores Butchers the shop changed ownership again, when two former-Smithfield market employees Mr Anthony Ross and Mr George Cawdry trading in partnership as R & C Butchers became the proprietors. This shop had one last throw of the butchers dice as Economy Meats, and floundered. Since the failure of this latest venture no other butchers have traded at this address. Other enterprises since have included a hairdresser, and a café in between long periods of vacancy. The shop premises are currently occupied by opticians using the name of this road within their trading title.

Figure 61: Price Brothers, Meat Stall, 1909.

Further along this road between Weedington Road and Carlton Street were shops retailing in Wines & Spirits, China, Toys and Boots. At number 102 sited on the first corner the retail butchers began there reign in 1893, beginning with Harry Hobbs. He was one of five retail butchers to occupy this site, albeit one Thomas Stone changing his trading name as the result of a dissolved partnership. Original Stone & Roberts from 1899 to 1905, he continued in business there and other sites after the break-up. The manager during this period was Thomas Barrow a twenty-four year old unmarried man from Swindon in Wiltshire. He lived above the shop as did two of his staff, Frederick Dicks a journeyman butcher and Herbert his younger brother. Both of these young men were born and bought up in Holloway, and were probably recruited via one of the many shops owned by the Stone family of butchers sited in the high streets of Holloway and Islington.

For the majority of its past life, this end of Queens Crescent has thronged with the vitality of costermongers, traders, shopkeepers and shoppers amid the congestion of stalls, barrows, horses, carts, and stock. In 1900 shop keeping families accounting for a sizable percentage of inhabitants in the market end of this road; often overlooked, many shop keeping families that served the communities shared the same social amenities and services as did their live-in employees', divided only by the fellowship of their trade. The public perception that all shopkeepers were wealthy individuals was a fallacy.

THE MARKET

The exact date on which Queens Crescent market opened is unknown, general consensus is sometime during the early 1860s. A grand opening of market stalls is unlikely to have happened, street traders appearing as each run of buildings were completed. Traditionally a food market from its inception, meat was among the various commodities for sale. Although contrary to popular myth, clothes were sold there from at least the 1880s and possibly before this date. Stalls were generally owned or rented by people with insufficient capitol to open a shop or used by established firms to expand their selling area. It can be argued at this point the last mentioned category is not a true market trader, but a shopkeeper.

The spirit of free enterprise is personified by the indomitable costermonger a name originally given to traders who sold 'Costards', a variety of large red apple grown since medieval times; monger meaning purely a trader or dealer. In direct lineage of costermongers are the Pearly Kings and Queens, known collectively as 'Pearlies', the aristocracy of street sellers. Supplies of fruit, vegetables and flowers were sourced from Covent Garden, Spitalfields and Kings Cross potato market opened in August 1868.

The wholesale markets used by past generations of local stallholders, among them Hetty Scott, Carters, Coopers, Blooms, Russells, Wheelers, Whymans, Felix & Sons, Butler and of late passing Albert Romain of 45 years standing. It is not widely known that street market workers, of which there were upward of 11,000 operating in London during the 1850s, had their own trade organisations; among them the Bethnal Green Costermongers Union, the Whitechapel and Spitalfields male and female Costermongers Association and the Street Sellers Union, later the Street Traders Union and National Association of Street Traders. The last named association had incurred substantial financial liabilities by 1929 and following a meeting at St Silas's Hall in Penton Street Finsbury was voluntarily wound up and liquidators appointed.

Prevalent in markets was the use of Back-Slang, a coded language reputed to originate from the days of highway robbers during the 17th century, as a form of crude slang to avoid detection. It was quickly taken up by elements of the lower criminal classes, many of whom frequented or worked in street markets or docksides, providing ideal locations for passing on stolen property. Genuine street traders adapted the language to suit their own particular trade, its use eventually becoming widespread throughout London. With regard to Butchers back-slang, it was perceived by those in the trade as fostering the notion of a mystic art or bonding process. In reality it performed a variety of other functions and was commonly used to deceive or speak unkindly about customers. Totally unintelligible to outsiders when mixed with trade nomenclature, it could be used in a positive way in large open fronted market shops to alert staff to shop pilfering without arousing the suspicion of the culprit. This proved useful during periods of rationing, when unguarded words meant a deluge of housewives descending on the shop, alternatively school boys were employed as lookouts. The multiple grocers positively encouraged the use of simple basic words such as ragus (sugar) and nocab (bacon), when anticipating delivery in order not to alert customers beforehand. There is a misconception due to the title back-slang that the letters in every word are simply reversed. On the contrary, the language is spoken at great speed using a combination of systems; it was rarely written down and of crucial importance the pronunciation of each word passed down to each generation of butchers. Its general use in the retail trade has declined with the reduction in the number of butchers shops, although occasionally phrases can still be heard spoken by the rehctub i.e. butcher.

Whether created by design or accident, Queens Crescent market was helpful in reducing obstructions caused by costers on other roads in Kentish Town, the title coster by then indiscriminately applied to all kinds of market traders. Many traders preferred their official title Barrow vendors or holders as used by local authorities.

The antipathy shown by some shopkeepers towards street traders was balanced by self interest, in the knowledge that their removal from Leather Lane market reduced the takings of shopkeepers there by half within a fortnight. The street-sellers were speedily allowed back after representation to the police authorities, as Henry Mayhew explained, "One reason why the shop keeper's trade co-exists with that of the street sellers, the poorer women who have to prepare dinner for their husbands, like to make one errand do. If the wife buys fish or vegetables in the street, as is generally done, she will at the same time, try to buy her piece of bacon or cheese at the cheesemongers, her small quantity of tea and sugar at the grocers, her firewood at the oilman's or her pound of beef or liver at the butchers. So one errand is sufficient to provide for the wants of the family". Although he was unaware, Mayhew

Figure 62: Bob's Butchers, 90 Queens Crescent, c1970.

was describing the principle behind one stop super-market shopping developed by a later generation of retailers.

Always more stalls than spaces, established shop retailers exacerbated the problem by using pitches which morally belonged to genuine street traders. Competition was fierce as pitches were on a first come basis when the bell and later a whistle sounded to open the market. In the early days squatters rights of the best pitches were nearly always secured by shopkeepers whose workers lived on site, leaving many costers to ply their trade where best they could. Options being limited, the vestry received complaints daily concerning errant costermongers causing nuisance by pitching in Kentish Town high street and Chalk Farm Road, and in 1871 warning notices were sent out.

Many of the costers chose to ignore the threat to their livelihood and further notices warned them to expect heavy fines. The Farringdon Road costermongers in the summer of 1892 were to be similarly harassed by the Holborn Board of Works. The obstruction committee empowered under the 1871 the street Clearance Act. The dubious reason given for eviction this time was unfair competition with the new vegetable market, the solidarity of opposition by costermongers the like of which had never been seen before in London. Their indignation culminating in a protest march by banner waving street traders from across the metropolis included five bands. The march followed by a meeting and further demonstrations at Clerkenwell Green, persuading the authorities to reverse their decision.

Whatever the desired effect of various pieces of legislation introduced, it failed to stamp out the problem; for nearly seventy years later itinerant costers were still active in nearly all the London Boroughs. In October 1900 the Hampstead Borough Council initiated a zero tolerance policy with a combination of fines and imprisonment and fifty-three costermongers, dealing in various commodities, were summoned to appear before the magistrates at Marylebone court.

In its heyday Queens Crescent market attracted hundreds of shoppers from far afield presenting a nuisance of a different kind, the absence of a public toilet. In 1895 construction work began to build an underground convenience at the entrance to the market which opened on February 18th the following year. Although welcomed by the shopping public its location in the middle of the road, with hardly a cart width either side, presented considerable difficulties for market traffic. The problem unresolved until the 1950s when it was re-sited parallel with the Sir Robert Peel public house; in line with the present day toilet masquerading as a black painted sentry box.

A few practitioners of the market trader's art, along with Sainsburys, became household names. The company history well documented in several publications, opened their second branch at 159 Queens Crescent and immediately extended their selling space with bacon stalls outside. With additional branches in other areas, business was phenomenally successful enabling the building of offices, stabling, warehouses and bacon smoking kilns in nearby Langford Mews in Langford Road, a side street leading off the market and later renamed Allcroft Road.

A polish immigrant, Michael Marks, started with a stall in Leeds open air market in 1884, a notice on his stall stated "Don't ask the price, it's a penny"; joined forces with Tom Spencer in 1886 to form the partnership of Marks & Spencer. Their Penny Bazaars were the forerunners of the stores we have grown accustomed to today, a branch shop in Kentish Town Road stayed for fifty years. The beginnings of Tesco are now legendry, Jack Cohen, with a stall and barrow in east London, joined forces with Mr T.E. Stockwell from whence the famous name derivation was formed using Stockwells initials T.E.S. coupled with the first two letters C.O. of Cohen, later

Sir Jack Cohen, he died in 1979. Already mentioned Messrs Lowe and Lea started with a meat stall in Tamworth market, this concern subsequently expanded to become Baxters (Butchers) Ltd.

Considerable sales of meat, poultry and game were generated through such markets. Information gathered from local councils used by the Royal Commission on Market Rights and Tolls in 1888, demonstrated that butchers meat accounted for a high percentage of the perishable foodstuffs sold. The officially recorded number of principle markets in 1901 for St Pancras totalled thirty-six. A section taken from the photograph of Queens Crescent market c1909 depicts a meat stall, which although details are largely obscured by its canopy is fairly representative. We can be reasonably certain, what are in fact two stalls combined belonged to the Price brothers, with a shop nearby at 54 Malden Road at this date. The main photograph also tells us the woman wearing the smock maybe tending the stall while the butcher is away, on what appears by the casting shadows to be late afternoon; in all probability during a weekday when the days trading had almost come to an end. Of further interest the large piece of meat is a forequarter of beef and the butcher's blocks are made from tree trunks. In addition to the Price brothers were the Seary brothers selling meat from their stalls further along the road. There were others selling poultry, game, rabbits and live day old chicks and live eels from the fishmongers stall.

Figure 63: Queens Crescent Market, c1970.

A later example is Bob's butchers meat stall outside his shop, the proprietor of both from the 1950s the late Mr Robert Ambridge, his ruddy complexion and manner suggestive of a country born man. In later years he moved to Witham Town in Essex and commuted to his shop via Smithfield Market. Anecdotal information has it, untrusting of banks and accountants he was known to carry large amounts of money in a black bag from which he paid the meat suppliers. Only at the very end of his life did he relent and engage an accountant. He died above his shop in the early 1990s. In 1967 Camden Council prohibited the sale of meat, poultry, fish, bread, confectionery and sweets in Camden's markets, effectively ending an age old tradition, a decision from which many local food markets have never fully recovered. The decision was reversed, albeit belatedly, with the exception of cooked food in June 2000.

SATURDAY NIGHTS

In the dark evenings stalls were lit by means of oil lamps then later hurricane lamps using paraffin, replaced eventually by electricity bulbs powered using small generators or supplied direct from a shop nearby. On weekends and festive occasions the lights burnt late into the night, as that most evocative aspect of market life began, the late night auctions.

As Austin Burnett, who lived in Queen Crescent as a young boy nearly seventy years ago, relates, 'One shop used to have an auction on Saturday nights between 11.00 pm and midnight. Most of the working class used to put off buying their weekly joint until the appointed hour. The butcher used to start out with a leg of lamb or a piece of beef or anything else, and add a pound of sausages as time went on to make the offer seem more attractive. He always said he would be in the workhouse with the bargains he was giving to ordinary people'.

Whilst late night auctions are generally attributed to lack of shop refrigeration, and the introduction of frozen meat The custom of late night shopping is much earlier and related to a time when most employees were not paid until the close of work on Saturday, which for many workers meant late at night. In itself a bar to any settled domesticity and compounded by the practice of paying men wages by money order on public houses, until stopped by William Cotton a past director of the Bank of England. He was a philanthropist and man of some ability; whilst there he invented an automatic machine capable of weighing 23 sovereigns a minute to an accuracy of ten-thousand parts of a grain.

One time resident of nearby Carlton Road, Mrs Emily Matthews then seventy-four, described her particular childhood memories in this extract from an interview for Camden New Journal in June 1984, 'All the old girls used to pile out of the pub at closing time and do their shopping. One butcher used to throw meat at you, bits of beef, and legs of lamb just to get rid of it. Mickleborough's fruit stall used to open all night on Christmas Eve as they prepared their display for the following morning'. Truthful in essence as these anecdotal stories of meat auctions are they must not be taken to imply butchers perpetually gave free meat away or indeed suffered huge losses as a consequence. Today we are all familiar with the dubious business antics of Del-Boy in the television series Fools and Horses, but in times past it was not always understood the gatherings at meat auctions were being strictly controlled.

Figure 64: Dennis Cole, 74 Queens Crescent, NW5, 2003.

The Saturday night meat auctions came into prominence with the import of increasing amounts of frozen meat from the 1880s, before mechanical refrigeration was widely available, which as already discussed in chapter four, meant by definition required a rapid sale; added to which other fresh meat that would not be in a saleable condition for the following weeks trade. Back then and into the late 1960s, it was normal business practice for market shops to be used as clearing houses by other non-market branches, or even another business where the wholesale cost had already been defrayed. The commonest ploy used was simply to increase the genuine mark-up price and deduct this amount during the auction; thereby the butcher was not out of pocket. The buyer was not being cheated; merely a healthier piece of meat of your own choice could have been purchased for the same money instead of the weeks leftovers. For much of the meat then auctioned caveat emptor! (Beware when you buy), especially in hot weather or as often was the case be prepared to use liberal amounts of vinegar. There is no doubt that some butchers applied the business adage 'your first loss is your least loss' and sold meat at below cost price, and at midnight when all reasonable hope of a sale had faded and in isolated cases, gave meat to customers free of charge. The widespread distribution of free meat was never standard practice, although cheaply made sausages were occasionally offered as enticement.

It is also worth mentioning that, with the exception of staff shortages and war time rationing, when fiddly coupons had to be removed from or pencilled out in ration books, the length of a butcher's shop queue was no accident either, the shorter the queue the more the butcher chatted the longer the queue the less he chatted; on the premise the busier you look the busier you will be, as a queue attracts the attention of the passing shoppers.

The vibrancy and atmosphere of those market days is vividly described in the words of the late Mr G L Evans, when aged fourteen commenced full time employment with Rayners Ltd at 74 Queens Crescent in 1920. Here he recounts memories from those days, 'The stalls with their displays of veg and fruit etc, then there was Hales the tripe dressers, hot baked sheep's heads, sets of brains, cowheel, Rayners four nights a week, from 5.30 to 7 pm, hot faggots 1d each, pease pudding, saveloys 1d each, pigs trotters 3d a feast for a few coppers. Saturdays you started at 6.30 am till 9 pm for breakfast you were given about ¾ lb of rump steak, you took some round to the coffee shop and had it cooked for 2d with the boss telling you, 'look sharp don't make a meal of it', wages 15 shillings a week. The market was like Kangaroo Street everybody jumping about. All meat shops had open fronts then, the law was to be covered each side later all behind glass. Friday and Saturday, the cheap jack so called, would attract the crowds at the other end of the market. Saturday nights after 7 pm you started knocking out meat at low prices, there were six butchers shops and two butchers stalls all getting a good living, one costermonger told me you don't have to have a good education to make money. When we had a display of mutton and lamb we allowed an old lady to stand alongside with her basket of mint, and all day long you would hear her say 'that large bunch of green mint mate one penny'.

The shop where Mr Evans was employed began its commercial life in 1867 as William Spiess, jeweller and watchmaker, with the appearance of the first market stalls. On the arrival of Rayner & King in 1874 began its hundred and twenty nine year continuity of service as a retail butcher's shop. On the 25th June 1934 Mrs Florence Caroline Rayner Ellis and Grace Emily Marsh exchanged contracts of lease with butcher Thomas W. Buttling and a new chapter began. The final incumbent Dennis Cole departed in September 2003, bringing to an end the era of the traditional butchers shop in Queens Crescents. Only time will reveal if this situation will be reversed.

MEATHIST MINIATURE

The Laws concerning hawkers and peddlers (50 Geo. III, c.41 and 6 Geo. IV. c80) treat them as identical occupations. However, strictly speaking the 'hawker' sells his goods by crying them through the streets in towns. The 'peddler' on the other hand travels through the country visiting houses to solicit private custom, cold calling. The majority of peddlers were Scotsmen, many of whom amassed considerable wealth until the coming of the railways which reduced the trade to insignificant numbers.

8

TOLERANCE AND TRADE

A considerable number of activities associated with the meat industry came under the designation offensive and noxious trades. The legitimacy of their function, whilst not in question but due to their locations in residential areas, primitive processing methods, poor hygiene and unsuitable buildings they presented a nuisance and health hazard to the public. Such unpleasant trades as slaughtering, tanning, tripe and soap boiling, tallow melting, gut scrapping, blood boiling and others dealing in bladders, grease and bones were a constant source of acrimony in populated areas.

Of gravest concern was slaughtering, an integral part of the butcher's craft for centuries. The very title 'butcher' was interpreted literally as meaning slaughterer of animals for food. The butchers of medieval London, highly concentrated in one area, were constantly in trouble. Ordnances and proclamations contained scathing references to the slaughter of animals causing an abominable stench that poisoned the air of the city, while blood flowed down the streets. Butchers repeatedly refused any attempt to regulate their trade, and were consequently reminded of orders relating to 'entrails and other filthy things of their craft'. Medieval street names: Stinking Lane, Sheep Lane, Butchers Row, Bulls Head Passage, Bladder Street, Flesh Shambles, Tanners Row, Rabbit Row, Rotten End, and Pudding Lane; the word pudding then meaning entrails of beasts, bear testimony to the horrific spectacle and manner in which the trade was often conducted.

The city authorities were not unsympathetic, and on March 12th 1343, the butchers of St Nicholas Shambles were granted permission to use waste ground in Seacoal Lane by the River Fleet for cleaning and disposal of offal. However the situation showed little improvement, and householders felt obliged to petition the King against dire nuisances created daily by butchers that resulted in an ordnance being issued in 1361 requiring all great beasts to be slaughtered at Stratford or Knightsbridge, outside the City. The Butcher's Guild were perfectly aware of public anxiety, only objecting to the many bad practices perpetuated by its members and others on the grounds of hygiene in the mistaken belief that all would be well if these matters were addressed. They steadfastly defended the butcher's inherent right to buy on the hoof and slaughter their own animals if they so desired.

In 1638 Smithfield cattle market was established under Royal Charter to the Corporation of London. The filthy and unsavoury conditions that existed there and animal cruelty until its closure 1855 have been graphically described many times over by observers of the day and need no repeating. On 24th November 1868 the cattle market was replaced by the newly erected London Central Meat Markets, for carcass meat and designed by Sir Horace Jones the City Architect. The site chosen for the new livestock market was Copenhagen Fields, Islington and named the Metropolitan Cattle Market; designed by James B. Bunning that together with the Foreign Cattle

Market at Deptford became the major channels for the UK meat distribution, both sites with public and private slaughtering facilities. The Metropolitan market for home grown cattle and the foreign cattle market as the name implies for imported cattle, thereby preventing contagious diseases being transmitted to our domestic herds.

Slaughtering as a separate trade was not generally recognised until the early eighteen hundreds, and these were trained butchers who specialised in this type of work. In some areas, as in York, they went under the name of the Pennyman, a butcher who used his slaughtering skills full time and that is how the situation would probably have remained had not the municipal abattoirs came into being, as it was there were only 2,634 official slaughtermen in England and Wales by 1911 against 155,732 butchers.

However until then the problem of butchers who worked from hoof to block were intractable, five hundred years of London history would pass and still the practice remained inviolate. By the second half of the 19th century, hardly a district within its boundaries was without the nuisance of slaughterhouses, and correspondingly throughout the entire United Kingdom.

A state of affairs hardly envisaged in 1796 when St Pancras Vestry commissioned land surveyor John Thompson to record the first topographical survey of Kentish Town. In the terrier book which accompanied the map of this survey he was able to say 'This map faithfully exhibits the centre surface of that delightful, luxuriant, and picturesque part of Middlesex within the parish of St Pancras'.

In retrospect his comments and the following observation a reminder of mans fallibility, for within fifty years this unspoiled part of Middlesex would be littered with the baggage of human habitation.

> ***The hay in the neighbourhood of Kentish Town and Hampstead is all cut and the fair inhabitants of those villages availing themselves of the favourable weather take their evening walks in the meadows.***
>
> June 1786

With the growth of London from 1750 the Guilds authoritative control diminished to be superseded by those of Government and City of London authorities, albeit the latter administration had no jurisdiction in Middlesex, which then incorporated St Pancras. The issuing of slaughterhouse licenses began in 1786 whereby there operation was conditional on the owner submitting a certificate from the church wardens and parish priest consenting to their application.

It is as well at this juncture to state that any attempt to give a full exposition of the relevant government legislature and local bye-laws associated with the operation of slaughterhouses, or the retail meat trade for that matter would be impractical. Other than to acquaint the reader with the fact the first attempt to bring all the strands together were incorporated in the Meat Regulations which came into force in April 1925. These covered slaughterhouses, stalls, shops and stores, and transport. Broadly speaking before this period, then as now, an enormous amount of civil and criminal law affected the meat trader which applied equally to other trades. A further complication arose in the matter of by-laws in that some regulations in the City of London differed from those in the rest of London, which in part differed from other areas of the country. From the beginning the practicalities of policing regulations proved difficult, not least because those charged with their enforcement were reluctant to allocate sufficient funds for inspection and control; in many instances deliberately obstructed by the opponents of change and those with vested interests. Their sphere of influence extended into national government and local authority politics in which they lived or had business interests.

At the monthly sanitary committee meeting for St Pancras, vestry members discussed and adjudicated on a range of issues concerning the local meat trade and other sanitary matters. Their deliberations, influenced by interim reports from the resident medical officer of health, although on occasions disregarding the advice and on others taking unilateral decisions on their own. A closer look at ground level soon reveals the enormity of the problem that faced the newly elected sanitary committee, not least the backlog of nuisances inherited from the old parish committees. From the beginning any thoughts of a swift conclusion was premature, as the meat industry were in the ascent and soon ranking with the most powerful industries in the country.

Dr THOMAS HILLIER

Of great importance in the improvement of sanitary conditions was the passing of Metropolis of Local Management Act of 1855, allowing the hotchpotch of parish committees duplicating and obstructing each other's work to be abolished and replaced by a newly elected body of 123 vestry members. As a consequence of those administrative changes the newly elected vestry appointed Dr Thomas Hillier the first Medical Officer of Health for St Pancras. The area under his administration was divided into six sub-divisions in ascending order: Regents Park, Tottenham Court, Grays Inn Lane, Somers Town, Camden Town and Kentish Town.

In his first annual report submitted to the Vestry for the year ending 1856, his findings must have made dire reading in what amounted to a damning indictment of the current environmental health conditions in the parish; couched in language tantamount to a castigation of the members themselves. For it contained astute and pertinent observations with a candour bordering on incivility; 'inadequate housing, drainage, roads and water supply were a disgrace, the gas works a monster nuisance and cow houses of manifest evil'.

The butchers and other noxious trades fared no better, all had been fully examined during the year and the slaughterhouses had been licensed under the Metropolitan Market Act, 1844, but in his view licenses should not have been granted to the majority of premises inspected. Stating bluntly 'It is impossible that slaughtering of animals can be carried on amongst a dense population without proving more or less injurious to the public'. He urged the Vestry to oppose the renewal of the licenses hereafter for those premises he considered unsuitable. Furthermore, he continued, 'Any butchers proved to be killing in unlicensed places should be prosecuted; three had been fined for this offence during the current year'. He was undoubtedly annoyed as despite his strong objections, one successful license applicant was using a cellar for slaughtering. As to the condition of the animals he was in no doubt that a large number of diseased cattle had been slaughtered and sold as healthy meat. Other offensive trades lambasted were gut spinning, bladder blowing and paunch cleaning, skin and bladder dealers.

An intriguing character despite his bluff exterior, he fought tenaciously throughout his term of office until his death in 1867, to improve the quality of life for all Pancratians. The vigour and professional competence he bought to the Herculean task confronting him was one that would have daunted the stoutest hearted of men. Medical officers of health would come and go, but the perennial problem of private urban slaughterhouses remained. Yet there was a glimmer of hope when magistrates, sitting in special session, decided that no slaughterhouse could be established or newly carried on within forty feet of any public way, or within fifty feet from any dwelling in accordance with the provisions of the Metropolitan Building Acts. While no action, other than cosmetic alterations, could be activated retrospectively against the current licensed slaughterhouses the decision would hopefully effectively prevent new applicants from setting-up in densely populated areas. The act had already been instrumental in the refusal of fifteen licence applications for places deemed unsuitable for slaughtering.

Undeterred another five butchers had tried their luck during the year and had been summoned and fined for using unlicensed premises. The St Pancras vestry were still receiving license applications of mind-boggling incredulity and more than once they were asked to sanction the use of cellars beneath occupied houses in crowded localities. One individual seemed surprised that his application to use a photographer's hut, with the only access up a flight of wooden steps through an inhabited house in close proximity was refused. The absurdity of the situation had much to do with a scramble to make money supplying food to a population that was increasing at an alarming rate and needed to be fed. A butcher could expect a reasonable living in what was predominately a food driven society and where acquisition of consumer durables was of lesser consideration. The very thrust of Dr Hillier's Sanitary fourth report in 1859, had undertones of impending catastrophe unless his warnings concerning insufficient inspectors were heeded, there were only two inspectors of nuisances.

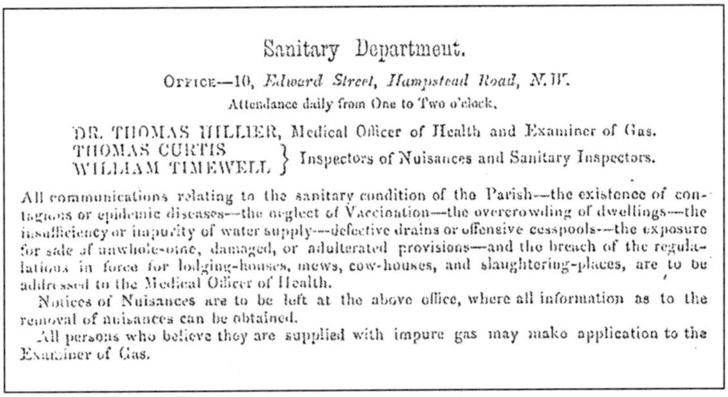

Sanitary Department.

OFFICE—10, *Edward Street, Hampstead Road, N.W.*

Attendance daily from One to Two o'clock.

DR. THOMAS HILLIER, Medical Officer of Health and Examiner of Gas.
THOMAS CURTIS } Inspectors of Nuisances and Sanitary Inspectors.
WILLIAM TIMEWELL }

All communications relating to the sanitary condition of the Parish—the existence of contagious or epidemic diseases—the neglect of Vaccination—the overcrowding of dwellings—the insufficiency or impurity of water supply—defective drains or offensive cesspools—the exposure for sale of unwholesome, damaged, or adulterated provisions—and the breach of the regulations in force for lodging-houses, mews, cow-houses, and slaughtering-places, are to be addressed to the Medical Officer of Health.

Notices of Nuisances are to be left at the above office, where all information as to the removal of nuisances can be obtained.

All persons who believe they are supplied with impure gas may make application to the Examiner of Gas.

Figure 65: St Pancras Vestry, public notice, 1862.

To emphasise the point he supplied the following statistics: The parish of St. Pancras covered 2,716 acres, was more than 16 miles in circumference, with a population of 200,000 and at least 21,000 inhabited houses. Their duties, in addition to investigating all complaints made to the Sanitary Committee, covered housing conditions, inspection of slaughter-houses, cow houses, all mews and wharves used to receive manure; visits to sausage-makers shops and street markets to see what kind of provisions are offered for sale and if necessary to seize articles unfit for sale and visits to magistrates courts and police stations.

In-door work such as keeping books, registers and time sheets added to their burden. It was indeed an unsatisfactory situation, with just two general inspectors responsible for such complex and diverse issues. The 104 slaughterhouses alone would require one inspector's full time attention. He would make repeated complaints on this issue during his term of office, with varying degrees of success. On one occasion the vestry refused, but commented the sanitary committee, were empowered to spend 4 guineas as assistance. Their working relationship was obviously fraught, with the vestry seizing on every opportunity to assert their authority, and at the same time justify their refusal to employ a reasonable level of sanitary staff; the ratepayers footing the bill could be relied upon to support the vestry. A golden opportunity presented itself when sanitary inspector Mr Curtis, was reported and not for the first time, for 'slowness of walking when out in the performance of his duties and for incivility to ratepayers'. Citing the imposition of gout in his defence, he received an admonishment for his slothfulness and off-hand manner whilst on duty. Another inspector, Mr Hartley, who was 'Spied standing for some time in conversation with his fellow inspector', was warned about his future conduct.

An altogether more serious matter was exposed in Marylebone Magistrates Court, with the corrupt behaviour of public health official Mr Scott of Kentish Town. As inspector of nuisances, he was charged on 6th August 1866 under the Cattle Plague Act, together with T. Sketchely butcher of Newgate Market. The charge was of fraudulently obtaining two cows valued at £44. 10s, the property of Mr Plant cow keeper of 31 High Street, Camden Town, pretending they were attacked with the cattle plague. This supposed guardian of public health was obviously taking advantage of the severe outbreak of the disease then sweeping the country involving over 324,000 cattle. After a two week adjournment both received their comeuppance in front of a court packed with dairymen, cow keepers and cattle salesmen when the inspector and the carcass butcher were each given 12 months imprisonment. Unfortunately, Mr Plant the cow keeper ended up out of pocket, receiving compensation to the value of only one half the worth of the two cows. The main defendant George Scott lived at 75 Carlton Road Kentish Town, having been appointed inspector of diseased cattle for St Pancras barely one year previous to his criminal activities.

By chance a year previous to this case and far from being an isolated example, another Kentish Town cowkeeeper, Charles Kilby of Rochester Mews was also duped. Having concern about one of his cows off its feed and fearful it might be afflicted with the cattle plague he sent it to the slaughterhouse of retailer and meat salesmen James Spencer, at number 6 High Street, Whitechapel. The instructions from Mr Kilby were that if his suspicions were correct the animal should to be sent to knackers boiling house or if sound to be sold for human consumption in the usual way. However the dressed carcass was found in a diseased condition among healthy meat on the premises. A fact in law that was enough for magistrate Mr Paget to fine butcher James Spencer forty shillings.

The wider events surrounding these cases are best explained by a report from Dr Hillier presented on 5th September 1865 to a meeting of the St Pancras Vestry: 'Since the last meeting of the committee the disease has spread to a great extent in the parish and 10 fresh cowkeepers have had the disease make its appearance in their sheds; making 23 in all since the commencement of the epidemic. The following are the names and addresses of the parties in whose sheds the disease has known to have made its appearance since this day fortnight, and the number sent away by each dairyman: Mrs Blyth, North Crescent Mews, 40 cows; Mrs Camp, 21 Argyle Street, 80 cows; Mr Tygate, 16 William Mews, 4 cows; Mr Cusden, Drummond Street, 5 cows; Mr Tucker, 22 William Street, 7 cows; Mr Pullen, 1 Guildford Street, 36 cows: Mr Kilby, Rochester Mews, 14 cows; Mr Greenaway 19 College Street, 7 cows; Mr Brown 1 Kentish Town Road, 26 cows; Mr Jelly, Highgate Road, doubtful number.

In all about 270 cases have been sent away in the fortnight and very many of these have sent cows before when manifest signs of the disease have shown themselves. It is very possible that other cowkeepers have suffered, inasmuch as they are anxious to keep the matter secret.' The number of cattle removed on account of the disease during the week previous to that referred to in the report was forty-four.

As previously mentioned the cowkeepers themselves were not adverse to unscrupulous behaviour when faced with a loss of earnings; many of them as in the case of inspector Mr Scott with the full connivance of market meat salesmen. The disease cattle plague, also known as Rinderpest, is caused by a virus which had ravaged Europe for fifteen centuries, the last occurring case in Great Britain happened in 1877 when it was finally eradicated. In 1894 the Diseases of Animals Act was introduced whereby previous regulations and inspection such as the Cattle Diseases Prevention Act 1866 were considerably improved.

Such were the rigorous enforcement of inspection procedures even before this time the mere presence of unwholesome meat on the premises justified a summons, be it shop or slaughterhouse; the offender, not always the guilty party as in the case of butcher James Hook of 317 Kentish Road. The case involved cattle dealer David Lloyd of nearby 27 Burghley Road who bought 300 sheep to send into the country and subsequently discovered 5 were unfit to travel. These he deposited at Hook's slaughterhouse situated in York Mews behind his shop, one of which was found to be diseased. The solicitor Mr Ricketts prosecuting for St Pancras Vestry argued both Hook and Lloyd were guilty whilst Mr Odgers defending stated Mr Hook never intended the animal should be sold for human consumption. The magistrate Mr Mansfield with regard to the hitherto unblemished record of James Hook and the defendant had not purchased the sheep dismissed the case against him, but found Lloyd guilty and fined him 40 shillings and costs.

In March 1869 further acrimony flared up between the St Pancras vestry and their employees, this time the Inspectors of Weights and Measures who were householders appointed under the local Act 59, George 111, cap 39. They openly accused the vestry of making efforts to keep secret the names of shopkeepers and others fined for using defective weights and measures. An article to this effect was published in the St Pancras Guardian newspaper wherein the inspectors made known, not only their intention to publicly name all offenders, but issued the first list of culprits. Among those named were two serving members of the St Pancras Vestry: Henry Honiball, butcher, 14 Queens Crescent, balance defective, fined 5/- ; George Harrington, greengrocer and chandler shopkeeper, 16 Queens Crescent, a potato scale and butter scale defective against purchaser and coal scale out of order, fined 7s . 6d and coal scale destroyed; John Evans, butcher, 82 Queen Crescent, balance defective, second offence, fined 1/- ; Luke Price Major, butcher, 147 Malden Road, balance defective, fined 10/- ; William Grey, baker 265 Kentish Town Road, balance defective, fined 10/- ; John Markham, pork butcher, 72 Queens Crescent, balance defective, fined 2s . 6d ; Robert Smith, 1 Phoenix Street, a piece of iron weighing two and three quarter ounces, under one scale, and two others defective, fined 15/- ; Wyatt, butcher, 13 Chapel Street, balance defective, fined 10/- ; Edward Browning, Chandler shopkeeper, 60 Seymour Street, balance defective, fined 5/- ; C. Blunt, coal dealer, 13 Charlton Street, coal balance defective, fined 5/- ; Barnett, eating house keeper, 66 Seymour Street, defective balance, fined 2s . 6d; Hannah Blake, chandler shopkeeper, 6 Little Drummond Street, balance defective, fined 5/- ; repeat offender and vestryman Mathew Knight, butcher at 208 Kentish Town Road was again summoned in February 1872.

A House of Commons return dated September 1868, for the number of persons convicted of using false weights and measures in the last six months of that year, in respect of St Pancras, Holborn, and Marylebone were 30 to 40 cases each. Pertinent to this subject it would be many years later at Marylebone Magistrates Court in September 1895, before the first case in which a person trading in St Pancras would be prosecuted for selling false weights. This dubious honour fell to a costermonger named Albert Rayner of Newbury Mews, Malden Road, Kentish Town, of whom it was stated had used ordinary iron one pound weights from which the inside had been skilfully drilled out and plugs of wood had been neatly fitted; these false weights known as 'ironclads' reduced by about fifty percent. In evidence Mr Bending inspector of weights and measures said his department had made considerable efforts over the years to trace the origin of false weights used by fraudulent tradesmen. The inspector explained on this occasion his assistant had been disguised as a costermonger to pursue enquiries. Costermonger Rayner was found guilty of having sold two false and unjust one pound 1lb weights and with having exposed for sale two diminished weights which were false and unjust. He was fined £10 pound plus cost, or undergo six weeks hard labour.

THOMAS STEVENSON, M.D.

With the appointment of Thomas Stevenson M.D. the new MOH for St Pancras there came a man no less irascible and diligent than his predecessor, to the extent of offering his resignation on one occasion. He drew attention to a Bill introduced into the Sessions of Parliament of 1872 for the perpetuation of private slaughterhouses; a bill that was subsequently defeated only to be reintroduced the following year. The bill was a pre-emptive move by the trade to prevent the enactment of clause s. 55 of the Metropolitan Building Act 1844, this by August 1874 would render it illegal to have any slaughterhouse within the metropolitan districts. Unable to hide his disappointment, Thomas blamed vested interest alone for delaying the time when these places legally ceased to exist. The butcher's trade societies had been instrumental in urging the public to support the bill, warning the abolition of private slaughterhouses would inevitably mean customers paying higher prices for their meat. For the time being progress towards eradicating the remaining 82 private slaughterhouses still operational in St Pancras had stalled. In the forlorn hope a compromise could be agreed, Dr Stevenson expressed his belief eight abattoirs; one sited in each ward of St Pancras and repeated in other parishes would be sufficient. The Metropolitan Police statistics for the year 1873 in which Kentish Town formed part of Highgate police division show the full extent of the problem that had to be confronted. Away from the confines of vestry meeting places, butchers still refused to accept this activity could not be tolerated in densely populated areas. The view of local butcher Robert Elvidge in Malden Road, on the Medical Officer of Health and his staff now renamed Sanitary Inspectors was robust. They were interfering with the legitimate conduct of his business affairs, an opinion shared by the majority of his fellow tradesmen.

The protest hardly surprising, given that in 1851 Kentish Town butchers together with Marylebone, Kensington, and City of London and even more so the Governors of Bartholomew's Hospital petitioned against the Smithfield Cattle-Market Removal Bill. Their opposition to change was based on the concern regarding additional costs incurred for centralised slaughtering, loss of personal supervision and disadvantages of transporting meat long distances. The latter concern was a reference to the still widely held view that the less meat is handled the better condition and keeping qualities it will retain.

THE OXEN AND THE BUTCHERS

"The oxen once upon a time sought to destroy the butchers, who practiced a trade destructive to their race. They assembled on a certain day to carry out their purpose, and sharpened their horns for the contest. But one of them was exceedingly old, for many a field had he ploughed, thus spoke: "These Butchers, it is true, slaughter us, but they do so with skilful hands, and no unnecessary pain. If we get rid of them, we shall fall into the hands of unskilful operators, and thus suffer a double death: for you may be assured, that though all Butchers should perish, yet will men never want beef."

There was however ample evidence to the contrary, Edinburgh, Cardiff, and Glasgow by the 1850s had all changed to centralised public slaughterhouses with no disadvantages to trade or customer. In fact the opposite, incidences of unsound meat had decreased and meat was actually cheaper than in Paris and Berlin where similar action had been taken by then with no adverse results. A host of legal enforcements including the Metropolitan Markets and Building Acts were still being used to try and prohibit the operation of private slaughterhouses; in densely populated areas with only moderate success, but changes were taking place. Adverse public opinion with the assistance of organisations like the London Abattoir Society, a modern day pressure group founded in 1883 for the centralisation of all slaughtering in London; the increased availability of frozen and chilled carcasses from the 1880's, together with many other factors meant the indiscriminate use of these places, hitherto an integral part of the Master Butcher's trade was becoming untenable. The inescapable realities of the situation still did not prevent the Butchers Guild from petitioning the London County Councils against the introduction of public slaughterhouses. In 1889, the London County Council took over from the Justices the registration and licensing functions and

POLICE DIVISION.	Number who have Private Slaughter-houses.	Number who buy Live Stock and have them slaughtered at Public Slaughter-houses.	Number who carry on their Trade, but do not Slaughter Animals.	TOTAL.
A. or Whitehall	1	- -	- -	1
B. ,, Westminster	53	15	52	120
C. ,, St. James's	16	19	36	71
D. ,, Marylebone	55	1	27	83
E. ,, Holborn	39	20	48	107
G. ,, Finsbury	62	7	42	113
H. ,, Whitechapel	36	- -	53	89
K. ,, Stepney	188	11	148	347
L. ,, Lambeth	50	2	25	77
M. ,, Southwark	48	3	50	101
N. ,, Islington	187	- -	81	268
P. ,, Camberwell	120	8	105	233
R. ,, Greenwich	138	9	79	226
S. ,, Hampstead	84	2	31	117
T. ,, Kensington	137	6	56	199
V. ,, Wandsworth	100	1	50	151
W. ,, Clapham	111	23	51	185
X. ,, Paddington	88	11	73	169
Y. ,, Highgate	174	22	59	255
TOTAL	1,687	162	1,063	2,912

Figure 66: Returns, Metropolitan Police Office, 4 Whitehall Place, 1873

the enforcement of legal requirements of slaughterhouses, knackers' yards, and cowhouses. In 1933, these powers were transferred to the Metropolitan Borough Councils, under powers conferred in Section 93 of the Metropolitan Management Amendment Act, 1862.

In a composite year based on vestry records from 1855 to 1900, two Kentish Town butchers, James Hatt of 11 Hawley Place and James Millar of 18 Malden Road were judged as unfit for slaughterhouse licenses for defective ventilation. A complaint was lodged against Mr Leach for boiling skin and fat in Charles Place, Grays Inn Road, and the vestry resolved to put down the nuisance under the Nuisance Removal Act.

The vestry also received a letter from Mr Field of 3 Mansfield Place, Kentish Town in which he complained of his next door neighbour, Mr Healy a butcher, was slaughtering in a small wash house attached to his residence that much to his annoyance was totally unfit for slaughtering and without doubt unlicensed. The outcome is unknown but the complainant asked for expenses as he was a witness against Mr Healey. The decision from the vestry was curt and unequivocal 'It has been resolved Mr Fields application be not entertained'. There were successful prosecutions against two butchers for killing animals in underground places, and Joseph Smout & Sons were refused a license for the same reason. In the case of Charles Daws of 23 Brewer Street cattle had to be taken down kitchen stairs and William Bright of 2 Charlotte Mews, Tottenham Court Road the animals had no water or ventilation. Three others Mr Holland, Charles Place, Kentish Town, Mr Charles Farey, King Street, Camden Town, and Mr Pursnol in Hawley Street, had been caught in the act of slaughtering animal in places not licensed for that purpose. The vestry were dismayed to hear that many others flouted the regulations and were therefore guilty of the same offence and resolved to oppose all licence applications for underground slaughtering. It is pertinent here to mention much of the slaughtering, legal or otherwise, was performed in semi darkness during the winter months because gas lighting very often did not extend to basements or back of shop areas; the work aided by the use of candle light or at best oil lamps.

Interim reports through the latter years showed inspectors had found a general good state in the slaughterhouses, although in some cases blood was thrown into the drains and ditches. In the early expansion of Kentish Town it was difficult to prevent this, as open ditches were frequently used to drain houses or for brick fields. The example of which Gloucester Place, Torriano Terrace, Winchester Place and Grapes Place in Holmes Road were cited; in particular Ferdinand Street, where open drains were the cause of a localised epidemic. The only solution then was the use of lime, or Mc Dougalls powder, for the most foul smelling drains and ditches. The inspectors found 161 Great College Street was being used for the slaughter of a large number of sheep and was filthy; proceedings against D. Nickles were taken. The premises of Harry Collingwood, Great Randolph Street, were deemed unfit for slaughtering anything but sheep, there being no pound for cattle and no ceiling.

With the arrival of railways supported by viaducts, the arches beneath attracted numerous trades and the butchers among them realising the potential for a slaughterhouse away from public thoroughfares. In Kentish Town providing a much needed slaughtering facility in 1868 for John Death, at number 1 the railway arches that sided Wilkin Street Mews and Frederick Turner in the railway arches that sided Grafton Road at the junction with the eastern end of Warden Road. In 1904 the North London Railway was associated as landlords of a slaughterhouse at No 6 the Arches, Priory Place, Camden Road; in 1939 renamed Prowse Place.

The public attitudes were sometimes baffling as shown by a petition received from Edward Watkins of 1 Cumberland Street, Goodge Street, in protest against the ruling that his slaughterhouse premises were a nuisance and unsatisfactory. He had used it for 26 years, and for 100 years before that it had been used by his predecessors. The local residents, it was stated, said that it is not a nuisance and he was willing to make the necessary alterations. Once butchers themselves realized these kinds of activities were unacceptable in urban situations a tolerable accommodation was reached with the local authorities. It was of course in their own interest because bad practices inevitably resulted in adverse publicity and consequently closure.

In 1899 the London Government Act came into force creating the Metropolitan Borough of St Pancras, councillors meeting for the first time on November 6th of that year. Three days later the Vestry Hall in St Pancras Way was renamed the Town Hall, the councillors inheriting a legacy of twenty-one licensed slaughterhouses; of these, sixteen in number were divided equally between Kentish Town and Camden Town. There were some progressive independent retail butchers, faced with abolition of private slaughterhouses that found openness was the best policy to accusations of malpractice and unhygienic conditions. A full page advertisement placed in the Hampstead Advertiser by Thomas G. Randall c1900, declared his entirely refurbished abattoir open to public inspection at any reasonable hour. One of the few licensed in Hampstead by the newly installed L.C.C., and it was claimed the most modern in the area.

In contrast one observer at this period wrote, 'Pigs with their throats cut were still to be seen in brine tubs behind the shop of Attkins the butcher'. The shop referred to was owned by George Attkins at 55 High Street, Highgate. Another unnamed observer commented 'Randalls the butcher has his slaughterhouse behind in the high street and cattle are driven up North Hill to it; the shop was sited at 82 High Street, Highgate, and run by Joseph G. Randall.

Figure 67: T. G. Randall, Abattoir, Elizabeth Mews, Englands Lane, Hampstead, c1900.

The references to the movement of herd animals, cattle and sheep, albeit the herding instinct of pigs is decidedly absent is one that often came to symbolise a time and place in the minds of the public, when the world around them was different than it is today. The herding of animals on public roads was a police matter regulated under the Metropolitan Streets Act, 1867. The general time limit within which cattle could be driven was between 10 o'clock in the morning and 7 o'clock in the evening, and in London the restrictions covered an area within a four mile radius of Charing Cross. In the period 1870-73, persons in the London metropolitan police district convicted of driving cattle outside the permitted hours totalled 663, and another 1,097 convicted for allowing cattle to stray in a public thoroughfare

The Metropolitan Cattle Market in Islington, with entrances in Caledonian and York Roads was opened by Prince Albert on the 13th June 1855. It had been designed and built, specifically with hygiene and efficiency in mind, at a cost of over £300,000 on land called Copenhagen Fields purchased by the Corporation of the City of London on which stood Copenhagen House together with a tea garden and cricket ground. Initially successful, with butchers who bought on the hoof, a little known earlier attempt by a Mr John Perkins to open a cattle market in Islington ended in failure. In 1835 he bought land between Essex Road (then Lower Road) and the Regents Canal and opened the following year; the venture lasted only a few months. From the onset officers of the Royal Society for the Prevention of Cruelty to Animals were on site to oversee the welfare of the animals. How many cases of cruelty took place and by whom during the lifetime of the market is unknown, except to cite one of many publicised prosecutions. The case in 1866 concerned George Irons a sixteen year old drover from Upper Victoria Road, Holloway, charged with gross cruelty. He was observed brutally beating several cattle with his drovers stick for no apparent reason. The Clerkenwell Magistrate, a Mr Cooke, said it was a clear case of gratuitous violence and sentenced the defendant to be imprisoned and kept to hard labour in the House of Correction for one month. In addition Mr Alexander, clerk of the market committee, stated it was a rule that if a drover was convicted of cruelty never to renew his licence.

For shepherds and drovers this applied equally to their dogs, in particular the drover where mistreatment was widespread and belatedly only addressed by the trade in 1892 with a show of drovers dogs. The first show of its kind held at the Metropolitan Cattle market to improve the breed of sheep dogs and persuade drovers to take more interest and care in their canine assistants. Among the exhibitors from the Collie Club a Mr Ballard of Highgate and Mr Henman from Breaknock Road.

Pleasingly the reverse of mistreatment to animals is recorded from regular meetings of the RSPCA (est-1824) where to mention one held in December 1872 at the Barnsbury Hall, Islington. The Baroness Burdett-Coutts (1814-1906) presented 64 prizes to shepherds, drovers, costermongers and cabmen, in recognition of their humane treatment to animals.

We have a window from December 1891 that offers another facet from a criminal case of robbery, with violence against James Gibbon a cattle dealer and drover who travelled from Leicestershire to Kentish Town Midland Railway cattle docks in order to tend his animals. On his arrival on the evening of Wednesday 28th he went to the drovers shed and sat down amongst a group of men and their dogs in the place where the drovers generally wait. Amongst them a man by the name of William Carter immediately began to provoke him by repeatedly asking if he wanted to buy a stick and stand him a round of drinks. Two other men named Taylor and Peterkin joined in the baiting and when he refused all three men violently attacked him causing severe injuries requiring hospital treatment. With the assistance of the cattle dock foreman and James Marwick a drover, who lived nearby at 156 Weedington Road, the victim James Gibbon was able to make his way to the police station and report the assault. At the trial all three men were found guilty and received sentences ranging from three to six months hard labour.

The principle thoroughfares leading to the Caledonian Cattle market in Islington were the busiest and caused the most problems. In Camden Road there was a permanent police presence with orders to caution drovers making more noise than was necessary in driving cattle or sheep, particularly in the early hours of Monday morning.

On the 10th February 1904, the Commissioner for the Metropolitan Police, by virtue of the Metropolitan Market Act 1857, issued new Regulations as to the Driving of Cattle with respect to driving cattle, calves, sheep, pigs and lambs and the control of persons driving or assisting, and also to the days and hours and the streets and roads by which they may be driven; a lengthy document covering railway goods yards at Kentish Town, Maiden Lane, Holloway, Battle Bridge and Junction Road, the route authorised to be travelled delivering or collecting animals on the days and times stipulated in the regulations, from which is taken the following example:.

Between the Metropolitan Cattle Market and the Midland Railway goods yard, Kentish Town, by way of North Road, York Road, Brecknock Road, Leighton Road, and Kentish Town Road, on Mondays and Thursdays between 10 am and 4 pm. Between the Midland Railway goods yard, Kentish Town and the Metropolitan Cattle Market, by way of Kentish Town Road, Leighton Road, Brecknock Road, York Road, and North Road, on Wednesdays and Saturdays between 3 pm and 7 pm.

The majority of the animals slaughtered in Metropolitan Cattle Market abattoirs were destined for Smithfield Market as carcase meat and delivered before the First World War by horse drawn wagons, a journey time of thirty- minutes. The wagons used in combination with the Foden and Sentinal steam vehicles. The sides of beef weighing on average 400 Ibs and requiring the combined efforts and technique of two men using a trotting motion (Smithfield Trot) to carry from the delivery wagon into the market.

By the mid nineteen-twenties droving of cattle, sheep and pigs by road between railway goods yards and the Metropolitan Cattle Market and other destinations was to be replaced by petrol driven vehicles. Although animals purchased by small and medium sized butchery concerns, with slaughterhouse facilities within a reasonable droving distance of the market, was commonly seen by the public, as retired farmer

Figure 68: Carcass Delivery Wagon, Smithfield, c1910

and butchers shop proprietor Philip Cramer in York Way, born in 1914 (chapter 10), and who recognises a herd of cattle when he sees them, confirms.

Also Mrs Reeks recalls the following scene in the early 1900s whilst out shopping with her mother 'Sheep were driven from Caledonian Market down Camden Road, up Park Street to Lidstone the butchers which was on the left hand side. The sheep went down a yard by the side to the back of the shop where they were slaughtered'.

An unknown contributor to the Camden History Society 1983 Essay competition writes 'One resident complained bitterly of a butcher, seen driving calves through his dwelling, to slaughter at the back. The butcher was unrepentant, saying if stopped he'd buy a place in the mews behind his shop and carry on the trade there and the result would be the same.'

The practice of droving animals still evident a quarter of a century later, witnessed by Reginald Pleeth, born in 1922, who lived at 55 Park Street, and spending his formative years in and around Camden Town. 'In those days, I do remember that once a week sheep were shepherded down Park Street, yes, it is true I tell you, and coaxed into the right-hand entrance of Lidstones, finally disappearing into the back of the large shop. At that time, I had noticed sheep grazing upon Primrose Hill, and it wasn't long before I put two and two together'. Some social historians have tended to dismiss theses claims as wishful thinking or parent indoctrination or some kind of thought transfer from one generation to another. In this instance the facts of the matter are tenders for the exclusive privilege of grazing sheep on Primrose Hill and other areas of Regents Park were still being advertised in 1921 by the Commissioners of his Majesty's Works, and anecdotal based information has it retail butcher George Thomas, Chetwynd Road, reputedly grazed sheep until the 1940s in York Way. The last verifiable case of droving in north Camden on an August morning in 1931 when a herd of pigs were photographed crossing Camden Road into York Road at the junction with Brecknock Road assisted by a policeman who obliged the drover by halting traffic. In countryside areas cattle and sheep were regularly driven through towns and villages to markets in the 1960s as witnessed by the writer. In London the droving of sheep is a right bestowed on the Freeman of the City of London and one that is still occasionally exercised.

Figure 69: Pigs in Camden Road August 1931.

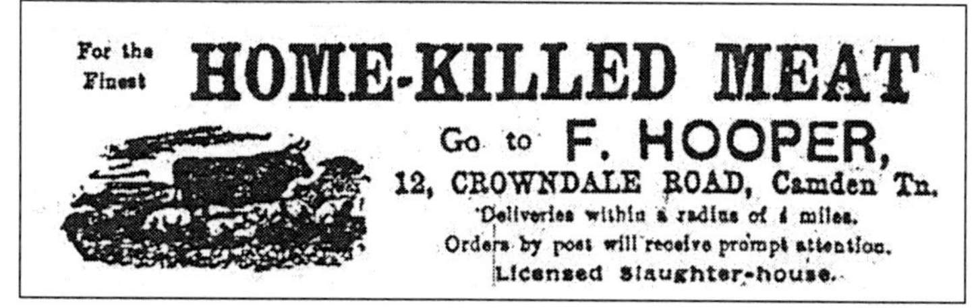

Figure 70: F. Hooper, Advertisement, 1912

As time went by regulations became ever more stringent in an effort to force the voluntary closure of slaughterhouses. In some instances restrictions were imposed, as with Lidstone Ltd, where large cattle were to be received before 8 a.m. at their

Park Street premises. In others, slaughtering was restricted to certain times during the day and so forth, while butcher F. Hooper, 12 Crowndale Road as was his right, still openly advertised in a 1912 edition of the Camden Town & Kentish Town Gazette, the provision of a slaughterhouse within his business. He was also subject to restrictions, this time the processing of small cattle only. This shop has traded under four butchers during its lifetime beginning with John Radden, Samuel Barrett, Frank Hooper and Frederick Toop. The family butchery business of the last named is believed to have been founded by William Toop from Reigate, Sussex, beginning with a shop in Caledonian Road and later 87 Parkway, Camden Town. By the time of the 1901 census 55 year old William had been joined in the business by his two sons 21 year old Frederick John and 22 year old Charles Toop, both born in Islington, Frederick it seems assuming ownership by 1915. The business was incorporated as a limited company in 1933 and still in business during the nineteen forties, however only the registration date and company number has survived from the official records. Mr Basil Leverton, of the Levertons funeral business, remembers Toop the butchers from his days as resident in Oakley Square nearby the Crowndale Road shop.

In 1927 there were over 20,000 slaughterhouses in England and Wales and at the outbreak of the Second World War some 11,500 were still operating. By 1953 these had been reduced to 482, of which 119 were public, 5 military and 358 privately owned.

MEATHIST MINIATURE

In 1849 Charles Fisher was appointed the first official Inspector of Meat in the City of London; the first Inspector of Live Cattle was Robert D Shouler appointed in 1851. A slaughterers Beef-Tree was used in abattoirs to hoist carcasses by being inserted through the achilles tendon. The implement was made of wood, hence Beef-Tree, from primitive times when animals were dressed outdoors and hung from trees, the carcasses stored in earth pits.

9

FIFTH QUARTER

Of considerable importance in relation to rearing animals for meat is the utilization and processing of edible and inedible by-products, which may roughly be divided as those derived from within and without. Very often regarded as unpleasant subject matter or dismissively referred to as trade spin-offs, the processing activities are by nature not given to public interest in this country. In comparison to America where no such inhibitions exist, Philip Armour, head of a leading packing house in Chicago, boasted in 1885 "I like to turn bristles, blood and the inside and outside of pigs and bullocks into revenue." If the reader will persevere, the brief inclusion here may help explain the more obscure trades and practices.

Beginning with human need for food, clothing, weapons, tools and shelter, for the greater part of our early history there was almost total reliance on what nature could provide; animals both wild and domesticated in particular were enormous providers of raw materials. From these austere beginnings a huge by-products industry has developed, from the useful everyday items to the frivolous, to the life saving pharmaceutical ones. The list and applications are extensive and from it can be traced the human reliance and long association with the animal kingdom.

By medieval times many of the manufacturing crafts earned their living from animal products, and indeed it was the reason for the very existence of many craft guilds, armourers, tanners, curriers, parchment makers, saddlers, harness makers, and cordwainers to name a few. A large proportion of the economic wealth of so-called merry England at this period was generated from such activities in the form of domestic and exports taxes. The merchants or their agents travelled throughout the country for the procurement of hides, wool and other commodities. There was a constant demand for hides, pelts, skins, wool, horns, hooves, and tallow for processing, to such an extent the value of meat from an animal carcass was relatively low compared to the profit margins of products it produced.

TANNING

The tanning trade was amongst the most financially rewarding, and consequently master tanners became some of the wealthiest citizens. Remunerations were well earned the work being physically arduous, dirty and foul smelling. There is no evidence so far that curing hides and skins was a major industry in Kentish Town. As with tallow melting, soap boiling, and tanning they were a minority industry; only Dixon, Jones, Taylor, Benson, Parker, Manne and Holmes, have thus far been recorded in the area. The earliest of those recorded, Robert Dixon, of whom unknown if in trade or premature retirement for the surrounding waters and air were considered to have health giving properties; whatever the reason his final days were lived in Kentish Town where he died in 1609. For an entirely different reason, tanners Peter Benson, Nicholas Parker and George Manne were indicted in March 1618 at Middlesex Sessions for assaulting and striking Richard Farrar in the King's highway.

Messer Jones & Taylor tanners are recorded as a partnership in Holden's 1805 Triennial Directory of London tradesmen and residents. However, their business activities remain undiscovered except for tanner C. E. Jones in Mansfield Place, Kentish Town, who made several appearances for bankruptcy between November 1815 and December 1819 at the Guildhall, London. Nevertheless it is conceivable Jones & Taylor may have local business connections from whence Tanhouse Field derives and now forms part of the land beneath present day Torriano Gardens Estate.

A name that appears repeatedly in early trade directories is William Holmes whose family sphere of operations included tanning, farming, brewers and house building. The name perpetuated in Holmes Road, so named in 1860 and which still exists today. Of interest here the tannery that with the benefit of a section taken from the map prepared by John Thompson in 1801, we know the process of tanning took place upon Tan Pill Field, bordering on Mansfield Place and Spring Place the former names of Holmes Road. The word Pill from the 17th century meaning to peel as in bark used in the tanning process. The field, triangular in shape, covered 1 acre 3 rods 17 perch, with tan house and pits together with office and garden. The outline of buildings and tanning pits are clearly visible on the map, and at this period the Spring Place section of the road crossed underneath by the River Fleet and running along the westward boundary.

The curing of hides needed a constant supply of water and in this respect the Holmes enterprise benefited from their close proximity to the river, accessed by a short path leading from the pits The throughput of hides and skins is unlikely to be discovered, although presumably neighbouring farmers and small holders took advantage of the service; as during the flaying process costly damage could result in unskilled hands. The task was essentially a manual occupation and enormous physical strength and stamina were required, cattle hides in the raw state averaging weights of 65lbs increasing to 80lbs for heavier breeds of cattle. The tanner and fellmonger trades encompassed all manner of animals both domesticated and wild, a side of Holmes tanning operations evident in the matter of a dog skin; in that during the autumn of 1799 at the Middlesex Sessions of the Peace, Thomas East servant to William Holmes was prosecuted for stealing a dog skin. The case versus East was delayed several times on the sworn affidavit of Richard Holmes that his father William was too ill to attend Court; he died several years later in April 1835.

The purpose of tanning is to convert raw animal skin into leather using mineral salts, tannic acid, and unimaginable today dung, including dogs' faeces that until around 1905 were still collected from the streets. The process of tanning remained unchanged for centuries, allowing us to reconstruct a hypothetical scenario of the method employed at the Holmes farm. Fresh hides, after cooling, would have been salted and dried. This first step which contrary to folklore performs no function in converting raw skin to leather the application merely to prevent putrefaction if the hides are stored or transported for any length of time. They would then be soaked in water to extract the salt and make the hides supple followed by immersion in a lime solution for up to a month to swell and loosen the hair. After removal of hair and dirt from the outside the residual fatty tissue and meat is scraped from the inside followed by thorough washing of the hide before it is transferred to other pits to be purged in water and dung. Thence transferred to the actual tanning pits in which a vegetable process of water and tree bark was used (tannic acid), preferably from the oak tree but other trees were used. The concentration of bark increased in the succeeding pits in which the skins were placed. Sheep pelts and pig skins were cured in the same manner, the sheep pelts referred to as wools prior to being shorn.

From start to finish the whole process would take three months or more, depending on weather conditions. Indisputably the nauseating odour wafting through the air pervading the nostrils on a warm summers evening and the possible pollution of the River Fleet at this point could hardly have endeared the Holmes family to nearby residents. It is thought a slaughterhouse existed in the vicinity belonging to the family. If true to what extent tanning work was derived from this source would depend on the throughput of animals from their own farm and if the slaughterhouse was engaged in work on behalf of other local stock farmers and retail butchers. We have no way of knowing the final destination of the hides or volume produced. It may have been used by local craftsmen or sold direct to London leather merchants. The bulk of the English trade was confined to cattle and sheep; the tanning of pig's skin took place mainly in the north and Scotland. By the 1880s, apart from a few firms situated near Smithfield the industry was centred in south east London.

By July 1819 the Holmes tannery had ceased operation and advertised for sale with immediate possession as a valuable leasehold estate, situate in Mansfield Place. The property consisting of 8 pits, 2 beam houses, 3 drying sheds, bark barn, mill, 3 stall stable and associated buildings, in addition supplied with excellent water, a dwelling house, garden and productive paddock. The following year, in the absence of buyers from the tanning industry forthcoming the land and buildings were offered to house builders, manufacturers and others wanting extensive premises. Four years later a family dispute interceded leading to 'Holmes against Holmes and others' in the High Court of Chancellery, wherein judgement was given that a considerable portion of the Holmes estate would be let by public auction at the George 1V public house in Spring Place, Kentish Town . The list of properties and land are too extensive to mention here, but included the entire row of buildings in Mansfield Crescent; the public house is still in existence situated on the corner of Holmes and Willes Roads.

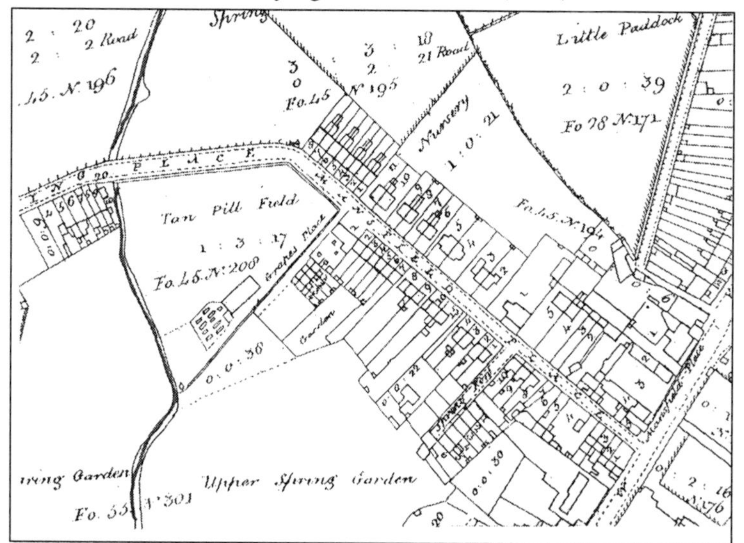

Figure 71: Tan Pill Field, section from Thompson's Map, 1801.

Allied to the leather trade were furriers, their guild The Skinners Company granted a charter in 1327, although they came together as a brotherhood much earlier in 1197. The company has long ceased to govern the craft and is restricted these days to administering their trusts and properties. A prolific benefactor of both in his time was guild member Sir Andrew Judd, whom upon his death bequeathed endowments of lands, schools and charitable institutions in London and Tonbridge Wells. In Camden his name survives in Judd Street and Medway Court an apartment block in association with his birth place in Tunbridge through which the river Medway flows.

TALLOW MELTERS

Of equal importance to leather was tallow and fat for candles and other industrial uses with demand for tallow; hard and soft fats nearly always exceeding supply. Tallow is a substance obtained by melting the harder and less fusible kinds of animal fats and incidentally responsible for the word pallor, or pale looking person. The London butchers and their wives were prohibited from selling tallow, suet and lard for the purpose of being taken out of the country. A total export ban on tallow was imposed on the city in the 15th century fearing a rise in the price of candles. Other towns restricted the sale of tallow to chandlers and candle makers only. Candles were made by dipping wicks into melted fat then cooled, the process repeated until the right thickness was obtained. Rush lights, a cheaper candle was made from thinly peeled rushes dipped to give a coating of tallow, and the most long lasting and expensive candles were made from beeswax. From the 1840s the whaling industry became a major supplier of oil obtained from whale blubber (sperm oil), used for lighting and other purposes.

The growing demand for candles, soap and fats ensured the ever present shops and factories, large and small, where this process was carried on within urban areas and much to the annoyance of the public and local health authorities. In May 1827 tallow melters Messrs Smith and Mills in Elm Street, Grays Inn Lane were indicted for establishing in January that year a nuisance in a populace neighbourhood.

A continual stream of witnesses for the prosecution voiced their objections; 'I cannot roast my Sunday joint without having it basted by the effluvia of the defendants tallow; None of the neighbours for a mile can open their windows, indeed every man that has a nose, passing by a tallow chandlers shop knows what a nuisance that is; the smell of tallow is very disagreeable particularly on Tuesday and Friday nights'. Despite the thirty witnesses for the prosecution sought by the local authority, expert witness in the construction of the steam boilers and furnace and an analytical chemist together with neighbours who said the smell was from the sewers then part of the Fleet ditch combined with smoke arising from the Coopers shop nearby. The jury returned a not guilty verdict on Smith and Mills, contrary to the obvious nuisance that only time and rationalization could solve.

To elaborate; the tallow melters of Tudor Place in 1842, the name of which conjures up a period in English history when candle lit castles and tallow boys were employed in the great kitchens to collect fat from animals used in cookery. The tallow melters were situated in a narrow turning at the southern end of Tottenham Court Road, in a patch work of buildings just above Hanway yard with stables extending back 300 feet, the party wall at the extremity forming the boundary of Black-Horse yard through which a thoroughfare existed for foot passengers to Rathbone-place. Here in Tudor Place was the extensive tallow chandlers business of Mr Morgan behind and within one door of his shop at 18 Tottenham Court Road; a thriving business of thirty or more years sited in this maze of twisting alleys and passage ways. The sixty year old proprietor employed forty local men, some of whom working day and night shifts in three warehouses engaged in tallow melting, soap boiling and making candles; the total throughput of 20 tons of soap and 80 tons of tallow requiring seven furnaces. Others employed as carters, drivers and stablemen collecting unprocessed tallow at night and delivering candles and soap to customers by day, the most prestigious being H.M. Queen Victoria. The density of this small area in Tudor Place evident from the following that occupied premises: Mr Jones also a tallow melter opposite Morgan; Charles Kerslake a tin manufacturer; Tenement buildings; C. Parker, cow keeper and St Patrick's Society School.

Adjunct to this story the transporting of tallow, as with sewage was then regulated by law to the hours of darkness. The last known tallow melters and soap boilers in this location were A. C. & A.H. Knight & Co, in 1901 with licensed premises at 1 and 6a Tudor Place, who are noted prior to this date in 1862, selling Tallow Grave candles, the melancholy named item presumably used for just that purpose.

Soap was comprised of lime, salt of vegetables and animal fats, and a method for making soap commonly practiced in the 18th century gave the following instructions 'A Lee or Lixivium is made of Kelp, that is salt of sea weed by burning, or of the white ashes of other vegetables, into which is added a quantity of lime water. When the Lee has stood long enough in the fats to extract all the salt from the ashes, it is then drained off and put into a boiler, with a proportion of tallow, (if for hard soap) or oil (if for soft soap), where it is allowed to boil until the tallow or oil is sufficiently incorporated with the strong Lee, and is become one thick consistency, it is then taken out with ladles and poured into chests, before it is cool they pour over it some blue, which penetrates the mass, when cold it is taken out of the chests, and cut into lengths with wire and laid up to dry'. Lime is a white caustic alkaline substance obtained by burning off limestone in kilns, for centuries a thriving industry in the Black Country area of the Midlands. Other uses included burning lime to produce an incandescent light on the theatre stage, Lime-Light, a phrase that has outlived the custom. Kelp is a large species of seaweed used in ash form which acts as a kind of soda.

In the early phase of personal hygiene awareness stood Cornishman Andrew Pears, a humble barber working in Soho, London and the founder of A & F Pears soap makers. He was by no means the first, but the originality of his transparent soap attracted considerable attention and the inevitable imitators. The soap was manufactured at factories located along the great Western Road, Isleworth; their place in advertising history and soap sales assured when they used 'Bubbles', a devastatingly enchanting painting to promote their product. The copyright of the famous Bubbles painting was bought for 2,000 guineas from Sir John Millais and was said to be a picture of his grandson and originally titled 'The Child', this angelic faced boy later became Admiral Sir William James. The chairman of the Pears firm from 1865 to 1914 was the eminent Hampstead historian Thomas J. Barratt and husband to the eldest daughter of Francis Pears; his inventive mind responsible for 'Bubbles' and Pears cyclopaedia. Midway between these two achievements and of greater importance historically it maybe suggested, lay his 'Annals of Hampstead' a painstakingly detailed 21 volume account of life in Hampstead published in 1912. The soap company eventually owned by Lever Brothers Ltd also published Pears Annual a magazine combining fictional stories, illustrations and advertising.

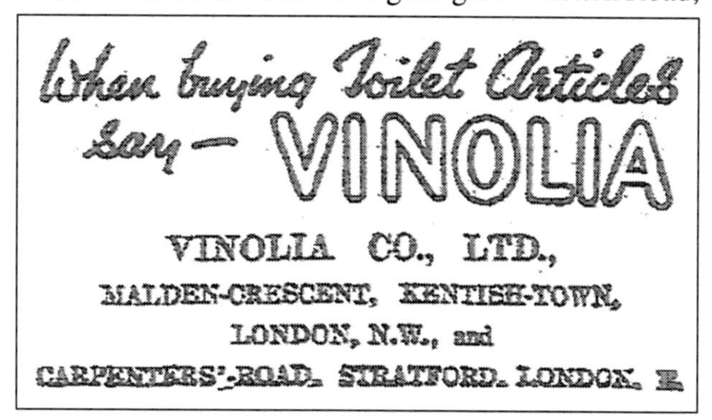

Figure 72: Vinolia Co Ltd, 1902.

Competition between advertising publications was fierce and in 1900 a new heavyweight magazine 'The Sphere' featuring in its time every product, from Bovril to Fry's chocolate, made an immediate impact with the public. Largely due to the editor Clement Short who had assembled an array of distinguished artists and writers to contribute their talents and of some, it was said, their artistic integrity, Anthony Trollope, Thomas Hardy, John E. Millais, Jerome K. Jerome, Sir Luke Fields, and W. Heath Robinson amongst them. From these, one time actor, teacher and clerk Jerome K. Jerome (1859-1927) with the middle name of Klapka lived for a spell at 15 Oakfield Crescent, Kentish Town. He found his niche in the lamp of fame as a dramatist, although today he is frequently remembered as author of Three Men in a Boat, published in 1889, one of several novels he penned. The house he once occupied was a local curiosity with the colloquial name 'the haunted house' used by parents to prevent their children trespassing, oblivious to nightmares resulting from the over imaginative minds of some children. The castellated building was enthusiastically impractical and worthy of Don Quixote. Towards the end of its life the mock ramparts fighting off only the ravages of time and indifference. The camouflaged appearance of gnarled paintwork and splashes of coloured rendering only drawing further attention to a visual deception until demolished during the Gospel Oak redevelopment programme.

That Kentish Town was briefly part of the soap industry has largely eluded local history books, the inclusion here because animal fat was then a major ingredient of soap and a by- product of the meat trade. The business involved were soap makers Vinolia founded in 1888, under the name Blondeau et Cie, until 1898 when it was converted into a private company as the Vinolia Company Ltd. Their first incursion into the area 1891-93 began with a warehouse in Ryland Road, for the distribution of a range of luxury soap, perfumes and preparations. In c1895 they acquired leasehold premises at Malden Works, Malden Crescent; a factory complex previously built in 1880 and commissioned by G. Rowney & Co, makers of artist materials and part used by them until 1967. From here occupying one section of the five story buildings Vinolia established registered offices, finishing, packaging and distribution centre. The rendering of soft fats and tallow and later vegetable matter was conducted at their Stratford factory northeast London, processing 690 tons per month of high grade fat shipped from Australia, New-Zealand and the Argentine. However beginning 27th of April 1899, the company convened a series of Extraordinary General Meeting at the Malden Crescent premises with a view to voluntarily dissolving the company for the purpose of reorganisation. All parties agreeing the re-registration of a new company, using the same name title, but with a different memorandum and articles was duly authorised. Worthy of note, the toilet soap world market place at this period was dominated by Vinolia, Gibbs and Pears the implication being Kentish Town was the beneficiary of considerable prestige and publicity.

In what had thus far been an eventful year, no less for the employees fearing unemployment than directors and shareholders, a further potential disaster on the afternoon of 26th August was narrowly averted. For on that day during a blistering London heat wave, with temperature officially recorded as reaching 90o degrees Fahrenheit, wood shavings and other combustible material lying on the factory roof of Rowney & Co, ignited causing a fire which also threatened the adjoining Vinolia premises. Fortunately despite the prompt arrival of several fire appliances only minimal assistance was required to extinguish the fire with the help of several employees. By 1906 the sales figures of Vinolia luxury soap and other toiletry preparations were attracting predators, among them William Lever who openly admitted would fair even better in more capable hands. A man of considerable determination, in that year he bought the company and within weeks expanded the advertising department, and vacancies for experienced pamphlet writers and display artist began to appear in national newspapers. Always years ahead of his rivals in business strategy his long term aim for Vinolia was the transfer of the business to Port Sunlight, Cheshire. A human industrial settlement he created for the manufacture of soap between the banks of the River Mersey and Birkenhead railway line. There followed not unsurprisingly in 1909, the news operations at the Vinolia works in Malden Crescent would be closed down; as time went on "Vinolia" luxury soap attained world status; supplied to first class passengers on R.M.S. Titanic and Olympic, both owned by the White Star Line. The product retained its popularity and is currently sold under the brand name of Pears Vinolia Soap.

Of reference to the Vinolia Works; at number 21 Malden Crescent, adjacent to the industrial site entrance was the birthplace of Stella Gibson,(1902-1989). A gifted writer, she is best remembered for her book 'Cold Comfort Farm' pub 1932, a comic portrayal of rural life. The daughter of Dr Gibbons the local doctor, she shared her early home life with the demands of her fathers profession until his sudden death on 15th October 1926, whilst treating a patient in his surgery. Of equal distinction and the opportunity to mention here; Kentish Town born novelist and travel writer Katharine Sarah Macquoid, with over sixty-five publications to her credit. Born on the 26th January 1824, Katharine was the third daughter of Mr Thomas Thomas a London merchant and was preceded by her death in June 1917 by husband Thomas Robert Macquoid, himself a distinguished artist and writer. Their unarguable talent in the fields of art and literature passed on to one of their two sons Percy Macquoid, a respected authority in the world of antiques and his brother a noted solicitor.

ABOUT THE FIFTH QUARTER

The fifth quarter derives its name from a body of beef or whole carcase when divided into four quarters and the contents within fall off or fall out when the carcass is opened, hence of-fal. From this imaginary fifth quarter, a wide range of edible foods are obtained some like tripe giving rise to specialist shops. The trade of tripe boiler or trippery was designated a noxious process and was indeed unpleasant.

The finished article considered either a delicacy or a horrible glutinous mess, depending on culinary tastes. Unusual on the menu today, tripe once provided a cheap and nourishing alternative to meat for the poor for a considerable period of its history and as with gruel and bread with beef dripping, nothing was more calculated to expose ones class status. In Edwardian times it acquired respectability when considered as a perfect breakfast meal, and positively health giving for nursing mothers, invalids and young babies. Yet today it remains an enigma, one of those dishes that was popular in previous times and still generates conversation today yet which later generations rarely know anything about. In truth, tons of tripe was sold fresh or vinegar pickled every year under a variety of names, grades and qualities to suit the most discerning palate. Of unknown origin, the word tripe is the name given to the edible parts of cattle and sheep stomachs, most commonly known were paunch or blanket, peck, kings head or honeycomb trip, considered among the best or bible tripe, red or black tripe, sheep's bag and rodican. All available from a specialist tripe dressing shop like William Cooper in the High Street Kentish Town, and Frederick Jones 143 Queens Crescent; both stocking a range of edible offal together with cow heel, udder, calves and sheep feet.

Certainly processing was a messy and time-consuming job, not recommended for persons of a delicate disposition. Broadly speaking the procedure was as follows; the stomach is emptied of its contents and turned inside out, then placed in a lime wash (lime and water), ensuring all parts are thoroughly agitated.

After most of the residual unpleasant matter has been removed it is then washed in cold water and hung up and scraped until clean. Immerse again in a vat of cold water and bring to the boil maintaining the temperature at boiling point for at least three hours. When tender soak the tripe in a bleaching mixture of water and soda or alum, finally wash again and store in iced water. The cattle and sheep stomachs were often sold raw and treated at home by previous generations less squeamish about these matters.

The trade persisted well into the nineteen- sixties in the north of England particularly in Lancashire and Yorkshire where one chain of restaurants reportedly sold 70 tons of tripe a week. One newspaper claiming the digestible qualities was the reason why North Country businessmen suffered less from ulcers than their southern counterpart.

Private slaughtering restrictions meant butchers eventually dispensed with processing their own tripe; instead buying in from high street tripe wholesalers. An important consideration here must be that these were bulk processes from the raw state bringing the premise under the scrutiny of the local health authorities and onto the list of noxious trades. From the late 19th century tinned tripe began to appear in grocery shops and was marketed by American canned meat firm Libby, MacNeill and Libby est., 1868. The tin can food process giving rise to the synonym "Sweet Fanny Adams" by English sailors, the contents of meat processed at Deptford cannery a connection between the macabre murder in 1867 of Fanny Adams in Alton, Hampshire.

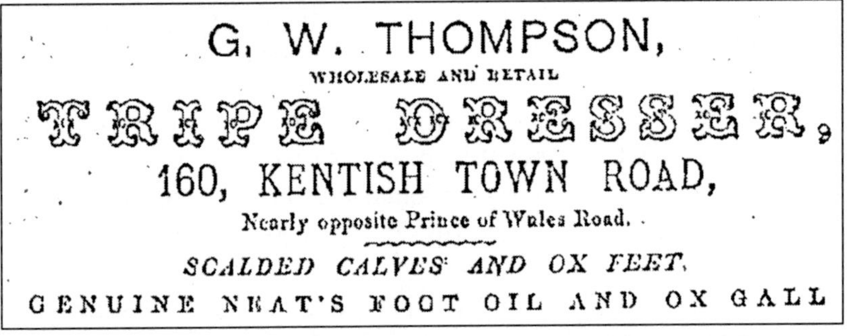

Figure 73: Camden & Kentish Town Directory, 1874.

Recorded at 160 Kentish Town Road from the 1860s George William Thompson, a Clerkenwell born man, living on the premises with wife Mary their five children and domestic servant Fanny Godfrey from Scotland. George came from a prominent and powerful family of tripe dressers based in Clerkenwell with shops in Islington, Norwood and Kent. In 1866 family member John Thompson, tripe dresser and builder living in Rosoman Street, Clerkenwell died leaving a freehold 3 acre estate in Dulwich, that by the terms of his will and codicil his legacy causing considerable turmoil within the Thompson dynasty. When finally settled the presiding judge commented 'An illustration of the penny wise and pound foolish, results which ensure when testators are unwise enough to employ non-professional men in the drafting of their will.' Much of the vast fortune left by the testator John Thompson was spent on legal costs and Kentish Town relation and tripe dresser George Thompson ended his days a discharged bankrupt.

From the Thompson advertisement will be noted the sale of other products and beginning with Neat's Foot Oil, this was produced by boiling down cattle feet to produce odourless oil used as a lubricant for delicate machinery. Although it was mainly used then in the manufacture of leather and is still used today by the equestrian fraternity. The use of the prefix 'genuine' on the container refers to adulterated oil made from sheep feet and other such like items. Also noted Ox Gall used in the manufacture of paints, dyes and a cleaning agent in leather. Other by- products such as bones were used for corsets, charcoal burning for engineering and as the name implies bone china. There were also bone handles for cutlery and weapon, the makers of the latter titled Knife Hafters. Before screw and flip tops were invented prepared skins (membranes) were used for covering jam, pickle and medicine bottles and a host of other containers. By the 1920s material derived from secretions extracted from animal glands were treating diabetics, thyroid and saliva deficiencies, snake bites, heart conditions, post traumatic shock following operations, mental illness and sexual neurasthenia, paralysis and women's disorders. Surgical ligature twine known as cat gut, but then made from sheep's intestines, and even gall stones used as charms or crushed for incense. Beef extract, fertilizers, cosmetics, gelatine, glue, ad infinitum.

Figure 74: Hales,Tripe Shop, 1904.

Well known for use in making black or white pudding, pig's blood was also used in the refinery of imported cane sugar to rid the raw material of impurities and fragments of cane. Tanks of crushed cane were mixed with ox-blood and when heated albumen from the blood carried the impurities to the surface where they were skimmed off. Nearly all of these ancient processes and some of the products produced from them have been confined to the back-waters of old technical trade books, but not all.

The next by-product 'Extract of Meat', sold as a nourishing drink for invalids when it was initially introduced, would have a futuristic role in food technology. The discovery of the beef meat extract, following extensive laboratory experiments by Professor Justus von Liebig, sold in powdered form to be diluted in water was produced in vast quantities at his factories in Fray Bentos, Uruguay. The public are more familiar with the retail trade names of Bovril, Bisto and Oxo, remarkably, intentional or not the first three letters of Bovine, Bison and Oxon.

The Bovril slogan 'The Glory of Man is his Strength' until recently printed on every jar. In a very short space of time it became clear this was no ordinary pick me up for the old and infirm and the potential for other applications such as stews, gravies and soups was quickly realised. The advantage of this concentrated food soon became obvious to explorers and military forces across the globe.

There were several attempts to enter this lucrative market using mutton essence, in this instance the brand name HIPI, needing some explanation to prospective customers by the producers G. Nelson, Dale & Co Ltd. who stated in their advertisements that HIPI was the native New Zealand word for sheep and should be pronounced "Heapy". In December 1995 butcher Harold George Wilson, aged 84 better known as the Oxo man died. He appeared in countless newspaper and magazine advertisements during the early 1960s; aside from his full-time career as proprietor of a butchers shop in the village of Great Bardfield, Essex.

CATS MEAT MAN

The sale of horseflesh for human consumption is viewed with disdain in this country and historically the British public have shown a strong dislike verging on total resistance to eating such meat; the horse being regarded as a noble creature and as with dogs and cats as companions. In ancient times eating horseflesh 'Hippophagy' was common in pagan religions, with the spread of Christianity the practice was outlawed and at times associated with health problems such as leprosy and blood poisoning. It will be of no surprise to readers therefore, to learn the number of local retail shops and dealers involved in this trade in modern times are negligible. However despite being a minority trade it was sold to the general public and is therefore a legitimate part of the meat industry and for this reason must be included. Open to malpractice from early on, an Act of Parliament was passed in 1889 to prevent the most obvious abuses. That laid down 'It is an offence to serve anyone with horseflesh who has asked for anything else' the bases of the regulations which still apply today. From contemporary reports misrepresentation and fraud were widespread and consequently unsuspecting customers may have unwittingly consumed horsemeat. In common with most other countries it is now mainly processed for the pet food industry or exported.

Figure 75: The Cats' Meat Man, 1884.

In Kentish Town as elsewhere most references to horse meat involve the 'cat's meat man', an early participant in the lucrative pet food trade. As part of the multitude of London street-sellers the 'cat and dog meat dealers or carriers', as they preferred to be called, numbered in excess of 1,000 in their best days. The meat, nearly always horse, and nearly always pre-cooked (boiled) at the knackers yard beforehand, was purchased and collected in a basket or barrow between the hours of 5 am to midday. Each dealer had their regular rounds, calling on customers on Mondays, Tuesdays and Saturdays, using the days in between for collecting outstanding debts and generally poaching new clients from fellow dealers. In the early period the trade was unregulated insofar as the employment conditions of juveniles was concerned.

From the premature death of John King a 14 year old deputy vendor of cats' meat and the subsequent inquest we have an insight of the trade. At the hearing held in August, 1844 at the Elephant and Castle public house in King's Road, (now St Pancras Way) it was stated the deceased suffered an epileptic fit whilst at the door of a customer in Henry Street, Hampstead Road and died instantly. Further evidence from his employer revealed the young boy was paid 2s.6p per week and his breakfast and dinner each day. At this juncture the coroner Mr T. Wakely, M.P. raised the possibility the laborious nature of the work and long distance travelled each day may have contributed to the lad's death. To which it was said a good cats' meat walk is a fine fortune selling at 3d and 4d per pound and he would not have to wheel in his barrow more at any time than half hundred weight; the jury returned a verdict of natural death.

A mention of celebrated cats meat seller in his day Henry Camplin born in 1829 and assuming the self mocking titles of 'Professor of Dogs Meat' and 'Purveyor of Horse Meat' trading and living in Wilsted Street, Somers Town. Following his death in December 1907 the business carried on by his son Robert

One startling observation at the time this trade flourished is the extent and depth of feeling shown by pet owners' themselves barely above subsistence level. A Kentish Town resident Mrs Ena Baker tells us 'We always had pets, the cats meat man called every day. He knocked on the door and if we were out he pushed a wooden skewer with little bits of raw horsemeat on it through the letterbox, and called the next day for his penny. In those days there was always plenty of horse meat available.'

Another local resident Mrs Ashbolt remembers, before the second world war, 'Along Pratt Street in a little turning called Pratt Mews there was a shop that sold cats meat, the man used to come round the street with a big basket and deliver the meat'.

In the same area, a stall in Inverness Street Market, Camden Town sold horsemeat for pets until the 1950's. Horse flesh dealer Robert Crook, of 39 Ferdinand Street, may well have had the ultimate business opportunity to supply the London Institution for Lost and Starving Cats and Dogs, situated opposite at 32 -34 in the same road. Carrying on the same trade at about this time was Harry Bradbury, dealer in horseflesh at 1 Newbury Mews, Malden Road. Dealers obtained their supplies from the knackers' yards at the Metropolitan Cattle Market via Lidstone Horse Slaughtering Co Ltd, which had premises there. The term knackers are thought to originally refer to harness or saddle makers in the 16th and 17th centuries. Others have suggested the term derived from the habit of harness makers having contractual arrangements to supply horses to the slaughter houses. This be as it may, from the 1900s knackers or knackers yard was in common usage as meaning a horse slaughterers yard.

The stark reality of this trade was although home consumption was limited, with a horse population of several millions; Britain was a major exporter to the continent. Largest of the licensed traders in the district was Harrison Barber & Company Ltd controlling six London and eight provincial depots from its head office railway depot at 180 York Way, Kings Cross; the firm originally John Harrison & Son until 1881 and thereafter forming a partnership with a Mr Barber. Morbidly, 'Dead Horse' was a slang term used by sailors in the days of the merchant sailing ships, meaning money owed to the boarding house masters on shore. The term also used in the lyrics of an old sea shanty 'They say my horse is dead and gone' meaning they had settled their debt. Our last stop concerns Arthur van Mingeroet of Dutch descent who opened for business at 167 Queens Crescent in 1953, the first and only horse meat shop in this road. It seems aside from the obvious financial aspect, his motives for the venture are unclear; presumably encouraged by the prospect of de-control of meat, whereby he could convert to traditional butchery or perhaps sell the business on as such. By 1958 the business had passed to butcher Mr D. King, a conventional retailer and then to Martin Stone, formally Martin Seelig, butcher and free lance meat buyer until in the 1960s the business floundering in a sea of financial difficulty.

MEATHIST MINIATURE

Such were conditions of factory workers engaged in gut scraping, washing and dressing of tripe and making black pudding and on account of the wet, unpleasant and offensive nature of the material processed; in 1919 the Home Secretary found it necessary to make the following order. Special provision is made for protective clothing, accommodation for overalls, and to eliminate the risk of infections of cuts and scratches to secure the welfare of the workers.

10

DIVIDED BY BOUNDARIES

The anomalies of shifting boundaries have, throughout the ages, divided and formalised hamlet, village, town, city, parish, borough and district; at times much to the annoyance of retail businessmen where shop location and prestige is of paramount importance. The shopper however concurs to no such system travelling where they please and for this reason are included in this chapter some additional butchers shops that skirt and cross the Kentish Town district boundaries, in this instance using the Kentish Town 1979 development map of Council Property.

To begin, the west boundary in Chalk Farm Road between 1831 and 1915; is a group comprising Weedon, Harbone, Bulpin, Walsingham, Wise, Frost and McCague. In the case of Weedon it appears he lived away from the shop for he is recorded in Pleasant Row, Camden Town, adjacent to the Britannia Public House. Although ten years on aged forty-three he had moved to Stuckly Terrace, Hampstead Road, without spouse but with three daughters recorded. At which time he had taken an insurance policy with the Sun Fire Office. The Sun Fire Office was established by Charles Povey in 1708 under the Exchange House Fire Office using the Sun symbol as its fire mark from whence it became known. This business formally constituted ten years later through a partnership known as the Sun Fire Office.

For Primest Joints and Cuts

you will be well advised in inspecting the stocks at—

G. W. CAGER'S,

35, & 37a, BRECKNOCK ROAD, N. 7.

PRIME ABERDEEN BEEF.

SOUTHDOWN MUTTON.

Pigs from Our Own Farms.

Figure 76: G. W. Cager, 1923.

From a later period, butcher Henry Walsingham at 31 Chalk Farm Road, presumably unrelated to Lord Walsingham, member and exhibitor of the Royal Agricultural Society, but with whom he shared a common interest. During the eighteen-eighties Henry, opened a branch shop at 50 St Georges Terrace, Regents Park; these references aside, in May 1878 Henry became a victim of a crime spree when his shop was burgled and money with other property was stolen. The act, perpetrated by William Morris, alias Coaly a labourer, Arthur Powell, alias Linton a butcher and gasfitter Edwin Lowther. The three men were all found guilty of breaking and entering the premises of several shops, including a draper in High Street, Camden Town.

Continuing north along Chalk Farm Road as far as Crogsland Road to a short section of Haverstock Hill that intersects with Prince of Wales Road and ends the west boundary, research thus far has not revealed the presence of any butchers shops. The meat shops along the northern extremities of Kentish Town along an imaginary line from Haverstock Hill across Parliament Fields east through Swains Lane to Dartmouth Park Hill, are dealt with elsewhere in the book, as is a short section of Camden High Street from Hawley Crescent to Camden Town underground station.

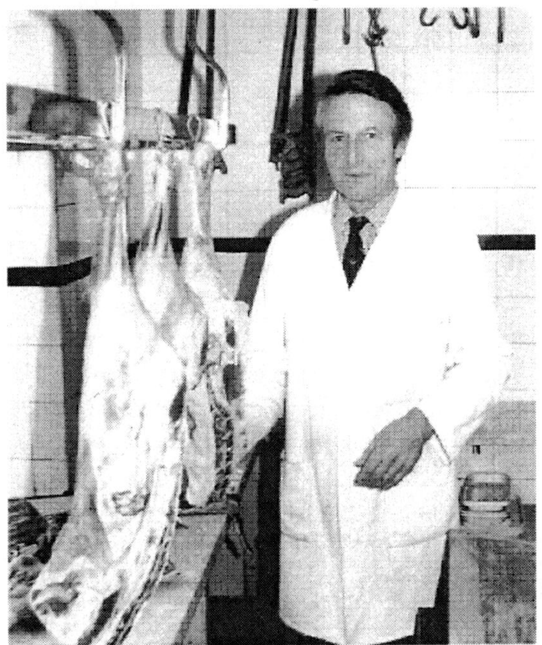

Figure 77: W. R. Mudd & Sons, 1985.

On the south boundary that runs along part of Camden Road and ends at Brecknock Road, only a few butchers opened for business the majority of buildings residential. By usage a shop at 63 Camden Road between Royal College Street and St Pancras Way were D. Doyle, Corrigan Bros and A. & .A. Whitton Ltd. The last named from c1937 to 1960, incorporated in August 1951 as part of multiple butchers John Manson Ltd until dissolved in January 1994.

Besides these and in recognition of long service were men like W R. Mudd & Son 25 Highgate High Street, 1955-1985, in the former Borough of St Pancras, and since 1965 officially in the Borough of Haringey, but certainly considered an integral part of Highgate Village by local residents. The founder's son Robin Mudd, seen pictured retired in the year this photograph appeared in the Hampstead & Highgate Express.

On the the eastern boundary in Dartmouth Park Hill were butchers: Joseph Bryant, George Ryan, Thomas Willis, William G. Harman, William G. W. Ivin, C. Elgar & Son Ltd, Mrs M. Kempton and Corrigan, the latter still maintaining a presence in this road, albeit under changed ownership. Of the aforementioned W. Harman with an additional shop in Torriano Avenue and A. T. Kempton a descendent of Mrs Kempton and still listed in 1969. Continuing into Brecknock Road which borders the Islington side and joins with Camden Road, the shops at number 18 and 35 in Brecknock Road are noted for continuity of trade in this road, thus far totalling nine proprietors between the 1850s until the 1990s, with the early buildings in this road with rooms and cellar below pavement level accessed by an external staircase and home to a variety of trades.

From the early arrivals like Henry Lee there in 1858, at Manville Terrace. He made application in that year to the vestry and was granted, a new slaughterhouse license for premises situated in Manville Yard with a rating of satisfactory, the site of present day Brecon Mews. Once again we may add the constant theme of longevity of trade within family owned businesses as descendant Henry Thomas Lee was still trading in 1913 from a butchers shop nearby at 116a Fortess Road. At number 35 Brecknock Road, Bedfordshire born Harry Barnes and St Pancras born wife Harriet, assisted by apprentice butcher Percy Coleby and later at the same address in the nineteen-twenties butcher George William Cager, from Brighton, Sussex, extending into a defunct upholstery business next door; the business by 1940 in the hands of butcher Cyril Isles.

This concludes the brief tour of butchers shops along the boundary's of Kentish Town with the exception of master butcher Philip Frederick Cramer, whose family business is not only sited at the crossroads of several contiguous district boundaries, but has been witness to much of the recent history in this locale and indeed the butchers community.

SOME ACCOUNT OF CRAMERS BUTCHERS

Cramers Butchers, at 392 York Way, Islington is approaching its eighty ninth successful year of trading and still in the same family. However the story begins in Nassau, a small town in the Lahn River Valley, Germany, where Friedrich Andreas Kramer the founder and father of Philip Cramer, the present owner was born. While still a young boy his mother died and his father unable to care for all his children, sort the help of relatives in England. It was decided young Friedrich should be sent to live with his uncle and aunt who owned a butchers shop in the Bethnal Green area of East London. The exact date is unknown but sometime during the mid 1880s he arrived in England to begin his new life, where he attended local schools in the area and immersed himself in the English way of life. As his son Phillip later recalled, betraying no trace of a German accent.

On leaving full time education, Friedrich joined his uncle and worked in the shop where he received a thorough grounding in the skills of the butcher's trade. It was during the summer of 1903 he married Ethel, a young lady from Norfolk, whom he had met in London, while she was in service in the town house belonging to the family who also owned Thickhorn Hall estate in her home village of Hethersett, Norfolk. On the 14th July 1908, under the Naturalisation Act, he took the oath of allegiance and became a British citizen and anglicised his surname. By 1910 he was in a substantial way of business on his own account at 36 Green Street, Bethnal Green, and within five years had acquired another shop in the same district at number 122 Globe Road. The rooms above the Green Street premises provide living accommodation for Frederick, Ethel and daughter Lottie and two German born employees; resident butchers Carl Leiblich and Frederick Stabb; in addition to William Pearce, a twenty-one year old butcher born in Custom House, an area in present day London Borough of Newham.

On the outbreak of the First World War he joined the Middlesex Regiment with whom he served until demobilised in 1919; his allegiance to his adopted country completely ignored in the prejudiced minds and violent action of certain sections of the public. A virulent hatred, that also prevailed throughout the meat trade with Smithfield market salesmen and Bummerees (porters) refusing to sell or handle meat purchased by German butchers. Many German nationals during this period were rounded-up and held in temporary camps, one at Alexandra Palace, until moved to an internment camp on the Isle of Wight.

The Cramer family suffering abuse and damage to their shops and boycotted by customers left them with no recourse than to close the business and seek sanctuary with relatives living in Hackney. The whole episode a harrowing experience long remembered by Philip, his parents, brother and two sisters. While away on war service, their mother became responsible for family matters and found accommodation in

Figure 78: Frederick Cramer, 392 York Way, Islington, c1920.

Goldsmith Road, Finchley. It was in this road he recalls being amongst an excited crowed of people watching a burning German Zepplin falling from the sky.

On his return to civilian life in 1919 his father purchased the York Way premises with accommodation above for the family. The former business of long established Mr Moore with a lease of twenty-five years, adjacent to a newsagent and confectioners, which separated them from the double open fronted shop of Lidstone (Butchers) Ltd, with slaughterhouse at the rear. A bold decision considering the presence of stiff competition from an established butchery firm, and the antipathy that still existed towards people of German origin in some quarters. The latter borne out by a close friend who confided to his son Philip that the previous owner had told him 'I have sold out to a German, but he won't last long'. However, there began for the Cramer family a long period of service and friendships in this road that continues today, which Philip remembers with fondness and wry humour, and in particular the adverse comments from a fellow trader.

His own formal education years began at Hungerford Road School before moving on to Holloway School in Hilldrop Road. During his formative years he well remembers the nearby Metropolitan Cattle Market at its peak and colloquially known as the Caledonian Market. 'The cattle, sheep and pigs being driven from Kentish Town caused excitement for us boys and some enjoyed the yelling and shouting , with the far from the gentlemanly drovers and dogs hustling the animals'. The nineteen twenties and early thirties, with the exception of the General Strike, was of establishment and improvement for both business and family.

Out went the old ice box in the backyard to be replaced by a new refrigerator in the cellar beneath the shop floor accessed by a trap door. The family enjoying for the first time the benefits of their labours with the purchase of a Bull Nosed Morris Oxford car allowing afternoon pleasure trips to Epping Forest. The vehicle garaged in Hungerford Road and the Cramers one of only two car owning families in the road, the other belonging to Mr Stanley Shaw at number 165, a successful grocer and proud owner of an Austin car, also employing a fleet of bicycle tradesmen delivering sausages and cooked meats in the area.

As the economic recession begun to worsen in 1929, causing high unemployment coupled with Philips disappointing school examination results that singularly failed to impress his father, he was grateful for the opportunity of a similar job delivering customers meat orders with fellow family employee Ron Daniels. One daily delivery he recalls in Goodge Street, being a considerable distance away a distinct disadvantage in bad weather, but also for the excessive weight of legs of mutton requiring him to sit astride the bike as a counter balance while the basket was being loaded to prevent the bike rising up at the back. Although he freely admits the perk of using the trade bike for more pleasurable pursuits, especially cycling to the running track at Parliament Hill Fields in preparation for the Islington and area sports day held at the sports ground at Tufnell Park each year.

It was at this period his training began in earnest to enable him to eventually succeed his father in the family business and no doubt having been raised in such an environment he was already well versed in the butchers way of life. When the prospect of accompanying his father to Smithfield Market in the early hours of the morning became a reality, he relished the opportunity. A journey by bus to Euston Road then by Circle line to Farringdon station and returning the same way, a journey that would become second nature. To the newcomer, the market seemingly a place of chaotic confusion, yet in reality the noise and hustle and bustle masking the buying and selling of meat; then an institution of immense importance to the economy of this country. The skill or art of buying meat on and off the hook (dead and live) requires a basic knowledge of arithmetic, long hours of practical experience and hopefully not too many mistakes; in the words of Philip 'As I was to learn many times at my cost. There was much to be learnt in being a buyer in Smithfield that could not be learned by reading a book'.

Once purchased, the meat was delivered to the Cramer shop by butchers carrier H.Todd, going by way of Caledonian Road to deliver to Freeman, Potter, Meissener and other butchers on route. On completion of business, father and son enjoyed a round of toast and a coffee at the Barley Mow pub before returning to the shop to begin a full day's work. For his part, midway in the nineteen - thirties Philip received £3. 10s wages for a week with half day Monday and Thursday; the time off providing an ideal opportunity during the summer months for afternoon family outings to their favourite location among the beautiful wooded area surrounding Asridge Park, near Berkhampstead. Convinced he had been an underachiever at school, real or imaginary, Philip enrolled for evening classes at the Hugh Middleton School for languages, near to Sadlers Wells.

A more permanent change of direction came through his close friendship with a cousin on his mother's side living in Norfolk, when he became involved with the country lifestyle through frequent visits to East Anglia. For it was because of this association he became the owner of one of the most unusual buildings used to accommodate a butchers shop. In 1937 he bought a butchers business from a Mr Park which was housed in a converted police station situate on the London Road in the market Town of Halesworth, Suffolk; a room that fronted the road and enabled a glass shop window to be installed, and a yard at the rear that police once used to escort prisoners to and fro now provided parking for the butchers van and delivery vehicles. The building with over sixteen rooms was obviously in excess of the average requirements for even the busiest butchers shop; but this was no ordinary shop premises as small goods production, a refrigerator and cleaning facilities were housed independently in three converted police cells.

Figure 79: Cramer Butchers, 2010.

Nevertheless the nominal rent of £1 a week proved an attractive business opportunity. The weekly shop meat throughput at the beginning he remembers as a side of beef, some pork and lamb direct with other requirements from the slaughterhouse of Mr Charlie Freeman and son in Beccles. In time he was able to supply and deliver meat, poultry and rabbits in excess of his own needs to his fathers shop, and also a butchers shop owned by his brother-in-law at 387 Caledonian Road. In turn he collected New Zealand lamb and other items ordered in advance from Smithfield market.

His affable disposition and enquiring mind it was not long before he immersed himself both in the social and agricultural aspect of country life, and eager to gain a thorough understanding of stock rearing, judging, buying meat on the hoof and slaughtering, working unpaid on many farms and in abattoirs to acquire practical experience and theoretical knowledge; his efforts fully rewarded on purchase and sale to his father of his first steer bought on the hoof, who commented seldom had he seen meat of such high quality.

His interest extended into poultry and game, and it was while purchasing Christmas poultry he was to meet Monica his future wife. Although their first encounter may have appeared less than promising, as Philip was the subject of a harmless practical joke; for Monica the young lady operating the weighing scale in a jocular fashion had her hand pressing down on the poultry increasing the weight to see if this Londoner had his wits about him. As future events proved, the young man was exceptionally astute and sold the Halesworth business back to the original owner. In 1940 Philip and Monica were married and settled back in London, although there was never any question of severing connections with Suffolk.

The return to his fathers shop in York Way found him in the thick of war time trading restrictions, although except for isolated cases there was to be no repeat of conditions suffered by the previous generation of German migrants. However, to the hard facts of the family business there were additional personal difficulties, one that affected thousands of men, the possibility of war service. Having registered in his age group, Philip was under no illusion when in June 1940 the buff coloured envelope stamped O.H.M.S. duly arrived. A worrying time for the Cramer family and their business for the duration of the war, the impact eased by super human effort from Monica and other family members, employers and close friends. To compound matters, while Philip was home on leave in January 1943 his father Frederick died aged sixty-three; a devastating blow that only reinforced their determination to keep the business going.

On return to civilian life and peace-time conditions after service in India, Philip and Monica were able to raise a family and plan for the future. To this end a small holding and stock rearing interests were entered into in Gosbeck, and in October 1950 the Cramer family moved to Suffolk; leaving the York Road shop in the capable hand of Ron Daniels. As time past Philip began to ponder the idea of acquiring a village butcher's shop, a convincing testimony that he missed the life of a retail butcher. Clearly his Suffolk venture had been successful and one he would retain and retire too later in life. There was of course a host of contributory personal and business considerations and not least family life, especially time spent with the children Paul and Jane.

In 1956 after much deliberation the family returned to London and settled in Muswell Hill. Happy to be in familiar surroundings and the working environment he knew so well, Philip began to expand the business. There was no deliberate plan, but as a businessman of high reputation and long standing friendships amongst fellow tradesmen, he would be given first refusal of any intent to dispose of a butchers business.

In the nineteen-sixties three shops were purchased, all from friends and all for reason of proprietor retirement. The first in Park Road, N8 a short distance from Philips house and within a year or two the opportunity was taken to buy the neighbouring shop of Mr Kalloway at 396 York Way, and finally the long established family business of Aubrey Guyer at 18 Brecknock Road. The then Mr & Mrs H. Guyer retiring to Cambridge and with whom Philip had, for many years, enjoyed a friendly business arrangement. Today Philip in his nineties lives in peaceful retirement in his beloved Suffolk and the shop he and his father devoted their working life is still open for business. The shop, now in the capable hands of manager Paul Langley, who is serving a new generation of customers, and of whom it must be said has done a remarkable job during a period of unprecedented change in the history of the retail meat trade.

MEATHIST MINIATURE

The anomalies of boundaries can in part be attributed to the introduction of the postal service who implemented a system of London postal districts in 1857 that with various amendments by 1917 evolved into the present number of sub-districts today. They were formed for the efficient operation of the post service, in areas around sorting offices and were not to signify the beginning or end of a borough or any other historical boundary. However they assumed in the minds of the public a social and class indicator.

11

SILENT SALESMAN

The most important sales asset to the retail food business is the provision of a display area, whereby potential customers could view goods on offer before buying. A simple and straightforward fact well understood by retailers and shoppers down through the ages; it is also manifestly apparent that window displays, dubbed the silent salesman, have always evoked strong feelings of nostalgia; child or adult, we all retain visual memories of that special display whether it be toys, sweets, butcher, baker or clothes shop.

The practice of displaying goods for sale, via an opening at the front of the shop, did not evolve until the middle ages; the early shops, invariably a dwelling house, which in the days before house numbering only distinguishable from its neighbour by a carved wooden sign. The sign usually indicated the nature of the trade that in the case of the butcher depicting a sheep, pig or oxen. The retail butcher laid out the various cuts of meat for inspection and purchase inside the building on trestles, tables or perhaps wooden boards, in whatever room considered most convenient.

As the art of selling progressed, the items were moved outside where a wooden bench or stall was often used for cutting or displaying meat; a 'Shammel', from which derives 'Shambles' as in the Shambles of St Nicholas, a collection of stalls and butchers shops in 12th century London. Guild members were forbidden to sell by candlelight, and a law was passed in the City of York that stalls and benches must be kept empty underneath, as a place of safety for children to hide from barrows and horse drawn traffic.

Later customers were served through openings at the front of the shop; the openings becoming display areas where customers were invited inside to exchange goods and money across a table upon which money was counted, the shop counter. When closed the shops were protected with a variety of wooden shutters, one design of double wooden shutters were devised, hinged top and bottom to open outwards, the lower shutter let down as a serving counter, the upper raised to provide shelter. Alternately a single lower shutter, their size taken to extremes in crowded London proved a nuisance, and a decree was issued in 1580 limiting the size to 2 feet 6 inches beyond the shop front. For security when the shop was closed, a bell attached to the shutters providing a simple but effective alarm system. Despite the introduction of glass, butchers continued to have open fronts, as glass was expensive and was generally considered unsuitable for the trade as it generated heat in hot weather; and that produced by early glass makers was thick and prone to defects known as 'Bulls Eyes'; later becoming fashionable as a decorative feature. In addition the glass was produced in small panes which needed wooden supports thereby blocking out most of the display area. As a general rule, butchers avoided renting or purchasing shops where a glass panelled window had already been installed, not least for tax reasons. The introduction of larger windows made from pressed glass from the seventeen-fifties onwards had only a marginal impact within the trade, until the introduction of rolled plate glass in the nineteen twenties.

A window tax, called with justification “a tax upon the light of heaven and the enjoyment of light and air”, had been in force since the reign of William III. One of the strangest and unpopular assessed taxes ever imposed in England, which over the years evolved into a complicated system of exemptions for various trades and situations; it has been often said where economic motives predominated over the rights of individuals Needless to state, during this imposition there were repeated unsuccessful applications to parliament for its abolition, and it eventually fell to the combined efforts of the vestries of St Pancras and Marylebone, after being narrowly defeated by three votes in 1850; they presented another Bill to parliament the following year. This time on 24th July 1851 their efforts met with success and the Window Tax Act was repealed.

EXCESSES

Figure 80: Brightwell, 6 Swiss Terrace, Belsize Road, 1892.

From the 1880’s, as the population continued to grow and demand for meat increased, competition between rival shop keepers became evermore keen None more so than butchers who began displaying outside vast quantities of great beast carcasses, cuts of meat of every description; offal, mounds of sausages, and other meat products. The carcasses carved with intricate herring bone patterns, using caul fat known as spiders’ web to add interest. The displays often reaching first floor level with make-shift trestles, tables, boards, barrels and boxes bought into service. Unlike their modern counterparts, the butchers in the Victorian age could never be accused of having more posters than meat. The sale poster called ‘Barkers’ from the fairground barker, to bark loudly as in dog and aided by the ubiquitous pavement ‘A’ boards.

Royal warrant holder Robert Brightwell & Son, 6 Swiss Terrace, Belsize Road, NW6, who apparently had a reputation for the excellence of their corned beef, was an everyday example; the mutton carcasses displayed knee high from pavement level. A native of Glenham, Suffolk; fifty year old Robert was a show judge well versed in animal husbandry and lived above the shop with wife Elizabeth and their two children. Originally established in 1866 in Henry Street, St Johns Wood, the business extended to several branch shops.

The poulterers and fishmongers did likewise, that together with the liberal use of sawdust, soiled clothing of the tradesmen, accumulated pavement waste and traffic congestion the thoroughfares resembled Smithfield Market. It soon became common for more trade to be carried on outside than inside the shop, a fact that had not gone unnoticed by local government health authorities, prompted into action by a barrage of complaints from a weary and long suffering public. Their patience at an end trying to negotiate their way through the tradesmen’s obstacle course strung out along the pavement; at the same time avoiding the hazards of dripping blood, slippery lumps of fat and sawdust. Shopkeepers in Chalk Farm Road were proceeded against by the St Pancras vestry in 1870, for displaying goods outside pleaded their treatment was unfair on the grounds that costermongers were tolerated, but the vestry were unmoved.

In William Aplin Coles, the undisguised show of bulk meat and limited hanging space had overruled common sense and both shop and private entrance have been obstructed. The area of display therefore may have been extended beyond normal practice for graphical considerations; although customer entrances were commonly festooned with poultry and meat, an aspect of shoppers' inconvenience accorded scant consideration until recent times. The business was begun by the Coles family c1860, with descending generations involved in several shops. The parent shop shown was open fronted with walls of white glazed tiles throughout and marble counters. In the early days the family produced nearly all its own beef and mutton from animals reared in fields adjacent to Adelaide Road. At 39 Belsize Road butcher Richard Coles, a descendent of the founder, remained until 1980 with the original shop sign W. Cole still visible.

Figure 81: W. Coles, 6 Upper Belsize Terrace, Hampstead, 1892.

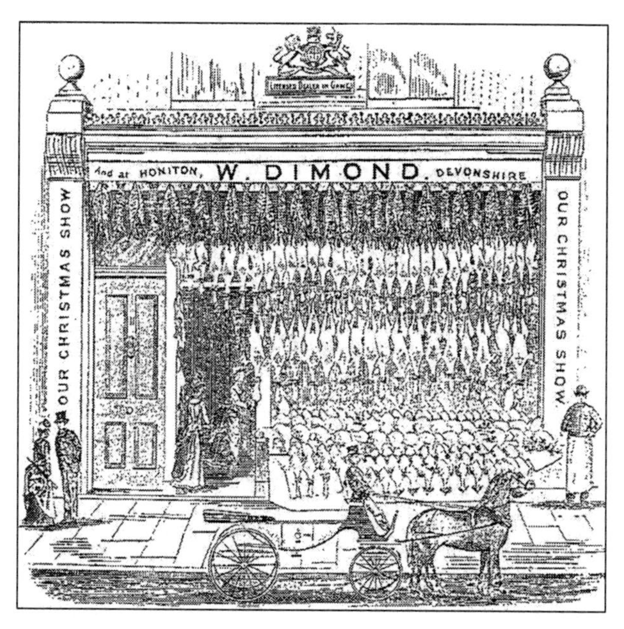

Figure 82: W. Dimond, 11 Fairhazel Gardens, Hampstead c1892.

Equally a more serious potential hazard was that of contamination from the dust, dogs, cats, horses and the germ laden atmosphere of the streets. After prolonged consultations between the various interested parties, recommendations for a change in the existing methods of meat purveying were accepted and implemented. In short, removal of meat from outside on the pavements back into the shop interior from whence it came, and thereby eliminate the worst cases of pavement encroachment. In effect, this only moderately reduced the risk of contamination, as the majority of butcher's shops were open fronted.

Of interest is a stylised drawing of a shop in Fairhazel Gardens, South Hampstead depicting a licensed dealer in game, poultry and general provisions. The display area festooned with all manner of birds to the extent of despoiling customers clothing as they enter the shop. The founder Hampstead vestryman, William Dimond, was born in Luppitt a village near Honiton, Devon where the family owned a farm supplying the retail side with dairy produce and consignments of poultry and when in season varied breeds of game. This particular shop was established c1800 together with adjacent oil shop, the family consisting of William, his wife Jemima and three sons living nearby at number 2 Coleridge Terrace with a nurse and general servant in addition to three lodgers. Of these James Griffin came from the same village as his employer and previously engaged as shopman at their Honiton branch. The other two lodgers W. Allen and Joseph Noyes were messengers, often as in this case titled letter carriers, a common source of employment before the wide spread use of the telephone. The resident retail butcher in Fairhazel Gardens at this period was T. W. Mavell sited opposite on the corner of Goldhurst Road; replaced at a later date by Duran & Levy, the partnership dissolved in June 1927 by mutual consent.

The introduction of glass panel shutters, proved a mixed blessing on account of their weight and fragile glass; so the trade breathed a sigh of relief with the introduction of lifting split sash windows, a typical example of which can be seen in the nineteen - forties photograph of Hammonds butchers shop.

The Hammond name is well known in Hampstead and this particular branch shop of the family owned by them in 1874 and at the turn of the 19th century consisting of master butcher Edgar Hammond his second wife Mary, butcher sons Frank and George, twelve year old Walter and daughters Fanny and Elizabeth. In the 1880s the adjacent shop at number 33 was a Fancy Drapers owned by a Miss H. Hammond. Also worthy of mention here, husband and wife historians John and Barbara Hammond authors of books on working class history amongst others the trilogy: The Village Labourer 1760-1832, (1911), The Town Labourer (1917) and The Skilled Labourer (1919). The cover picture of the first mentioned book, reprinted in 1987, instantly recognisable as Hampstead Heath painted by John Constable.

Figure 83: Hammond, 32 Heath Street Hampstead, 1942

A commemorative plaque for these eminent social historians in situ at Hollycot, Vale of Health NW3.

Although a tremendous improvement, glass panel shutters did have one disadvantage inherent in the design; only half of the window display could properly be seen at one time. Not until the fixed, mass produced sheet glass window, later toughened and approved by the health authorities, were the technical problems finally solved. It would take a statuary requirement in the 1920s for butcher's shops to install glass windows, and even this was only partially successful due to exclusion clauses. The butchers continued to display high volumes of meat, and poultry, regardless of glass windows or open fronted shops. Many of these sights are within living memory, while others are recorded in photographs, illustrations and personal testimonies of shoppers.

PAGE STORES LTD

We are fortunate in having an eye witness account of such displays, observed over seventy years ago, providing a wonderful historical cameo. Reginald Pleeth recounted his enduring memories of those impressive displays in an essay competition submitted to the Camden History Society in 1980. Reginald was born at 55 Park Street in 1922, now Parkway, where he lived until 1935. In this extract, he gives a florid description of Christmas time at Page Stores and Lidstone butchers in Camden Town.

'A butcher's shop window crammed with rows of well scrubbed pink pigs, bellies split open ever so neatly; in other windows not to be outdone though obviously 'done in' were ranks of sleek looking shapely sheep carcasses. These flaunted frilly paper collars round their necks - de rigueur in Camden Town butchers shops at that time it would seem. In particular, I recall Pages butchers shop, rather like a miniature Smithfield Market, with salesmen auctioning poultry and vast joints of meat outside the shop, while carrying on a stream of good-natured banter with the eager customers. There as always I averted my eyes when we passed the rows of skinned rabbits, complete with heads and lack-lustre eyes, displayed outside on a large marble slab. Finally, before going in doors, we re-crossed the road to look into the bright lit windows of Lidstone butchers right opposite where we lived. There one could see chains of sausages twisted into shapes of symmetrical perfection, circles of black pudding, piles of tripe and masses of joints and poultry arranged with geometrical precision. The focal point of this exotic tableau a boars head complete with lemon in mouth, displayed upon a large salver addition to the sheep and pig carcasses.'

It should be noted that for display purposes butchers invariably used uncooked domestic pig heads which could be utilised after, as wild boar were expensive and scarce. The genuine cooked boars head with tusks carved from bone was usually stuffed with a variety of force-meats which might include truffles, tongue and nuts etc, then glazed and decorated with piped lard or gelatine shapes, after glass dolls eyes had been inserted. For customers on a small budget and even smaller appetites a pate of boars head in glass moulds at a shilling each were readily available from upmarket butchers and grocery stores.

The shop referred to as Page's belonged to Henry Page, a pork and beef butcher established in 1848, and also proprietor of 231 High Street, Camden Town and Westmoreland Street, Walworth. His family, later trading as Page Bros, acquired additional premises either side of his original shop adjacent with Quality Stores, dominated meat and poultry sales in the high street. Two points of interest, rump steak displayed on the two draw cabinet in the centre of the picture is priced at one shilling and one penny per pound, and secondly the shadowy figures of customers and hint of a workman that can just be detected at the top right hand side of the ladder is repairing the wall rendering; these ghost like appearances a common fault in photographs taken at this period. On the 17th August 1950, the business was incorporated as Pages Stores Ltd., with registered offices above the shop of butcher Davis H. Churchouse est. 1913 at 67 Kentish Town Road, it is unknown if there were any business connections between them; although the fact of Henry Page's long association with this road, having traded from a grocers shop back in the 1860s, may suggest he owned the property. Page's Stores Ltd were dissolved on 10th May 1981.

The scene witnessed by Reginald Pleeth, with the exception of a Wimpy bar, differed little some thirty years later. For this one section of road from Inverness Street to Jamestown Road had a complement of butchers enough to satisfy the most rapacious carnivores. As huddled together with the two large butchers shops already mentioned, were multiple butcher John Best dissolved in 1971 and family butcher Edward Harrison together with Talby the fishmonger and Woods the greengrocer.

A hive of activity as closing time approached on Saturday afternoons, when butchers blocks and various items of equipment were taken outside to be cleaned. Passing pedestrians trod the slippery wet pavements and negotiated the buckets of boiling soda water at their peril; whilst only the most intrepid of customers entered the shops. A frenzied spectacle, as memorable as it would have been familiar to previous generations, and more than once private prosecutions were instigated against traders by members of the public claiming compensation for injuries sustained caused by the butchers cleaning-up schedule.

Figure 84: Page Bro's, 229-231 High Street, Camden Town, c1900.

Always open to abuse, among the genuine claims were the bogus perpetrated by unscrupulous individuals who saw a chance for easy money. A widely publicised case in December 1911 concerned Henry Page himself and another butcher John Edwards in Kentish Town Road. Two victims of a string of fraudulent claims made by costermonger Samuel Harman and associate Ernest Pendrigh. At their trial both these men were charged with conspiring to obtain money by false pretences from various named businesses. In court it was stated one or other of the defendants would pretend to have slipped upon a piece of fat outside the shop, then went inside and demanded a sum of money as compensation. Both Page and Edwards as members of the association of butchers had free access to the legal services. Their solicitor, Mr L. Ricketts, had little work to do as Harman pleaded guilty for this and other offences and received twelve months hard labour.

His partner in crime, Pendrigh was already undergoing eighteen months imprisonment having been sentenced at an earlier trial.

Returning to the butchers shop display the inclusion of a boars head was much favoured, although most butchers were unaware of its historical association with the trade. It appears, on company arms granted in 1541, a reminder of the Guild's ancient heritage during the reign of Edward III, when city authorities claimed annual rent of a boars head from the butchers of St Nicholas Shambles in 1343 for a grant of land adjacent to the Fleet River where they could dispose of animal waste. After a considerable time lapse, in 1547 it was decided to resurrect the boars head custom, by presenting the incumbent Lord Mayor with an entire boar, costing 24 shillings and signifying respect and obedience from the guild; the tradition continued until the end of the seventeenth century.

Figure 85: Boars Heads, G. Brazil & Co Ltd.

In the early 1970s the custom was revived and moderated and students attending Smithfield College were given the opportunity to prepare a boars head under the supervision of a tutor. With the closure of Smithfield College the privilege passed to the Smithfield Department, Walthamstow College until of latter years the preparation of the boars head being undertaken by trainees of the Combined Services Catering School, Aldershot. An annual event in November, the boars head is carried on a shoulder tray by students through the City of London streets and presented to the incoming Lord Mayor at the Mansion House. In consideration of hygiene a replica fibre glass model is used in the parade, whilst the edible boars head is delivered by refrigerated vehicle. One of the last occasions boars heads, probably domesticated, were consumed in any appreciable quantities was at the opening ceremony of Smithfield Market on the 24th November 1868. The function was attended by the Lord Mayor and 1,200 guests who enjoyed a banquet of boar heads and barons of beef. The true origins of the boars head feasts date from the rituals of the winter equinox practised by our Nordic ancestors, when an entire wild boar was offered as sacrifice to the Goddess Freia to ensure the restoration to vigour of the life giving sun.

With the spread of Christianity, the ritual was changed from one of sacrifice to one of feasting and by the end of the seventeenth century, wild boars were virtually extinct in England and most of the grand feasts lapsed. A version of these rituals still survives today on the occasion of the Boars Head Feast at Queen's College, Oxford, culminating in the singing of the Boars Head Carol.

The festive seasons of Christmas, Easter and other national celebrations provided an ideal opportunity for butchers shop displays. Although a common enough sight throughout the colder months of the year, stock was drastically reduced in summer. Fear of competition did in part force butchers into adopting the 'If they cannot see it, they will not buy it" attitude, resulting in these spectacular displays. There was also another factor, less apparent and more mundane, and that was inadequate storage space inside, an inherent problem in early shops. The introduction of ice boxes and later mechanical refrigeration did little to alleviate the problem for they in turn required a considerable amount of space, hence the reason they were located either outside the premises in the back yard or underneath the shop. Meanwhile, the sales area substituted for the general lack of storage space, providing a convenient place for sides of beef, mutton, lamb and pork carcasses until broken down into primal cuts, for with the exception of rural areas the sale of an entire carcass was highly unlikely before the days of home freezers.

Apart from window displays the majority of butchers were unconvinced of the power of advertising, complacent in the knowledge it was unthinkable not to buy meat. Practical men, unfettered with modern advertising ideology such as the function of advertising is to create a desire for goods and so forth. They believed people had to eat meat and that was that, why waste money repeating the obvious. The large volume of meat required during Christmas week, meant displays were often left out overnight, a friendly night watchman from the nearest road repair gang providing security and receiving a present of meat in lieu of payment. It was also the custom in days gone by at this time of year to exhibit live cattle, sheep and pigs either tethered in the shop or outside for customers to view, particularly where slaughtering facilities were readily available.

Figure 86: P.N. Brazil , 172 High Street, Camden Town, c1906.

P. N. BRAZIL

The scenario of meat displays quite familiar to older generations, judging from the childhood memories recalled by eighty-two year old Ena Baker, whose prize winning essay 'Shopping in Kentish Town' appeared in the 1984 issue of the Camden History Review. Ena lived in Ingestre Road and in this short extract she explains, 'Brassells the butchers in Kentish Town Road had the best quality meat. If you wanted a joint of meat for Sunday my mother would put me in the pram and wheel me to Brassells quite late on Saturday night. There were no fridges, so we got a bargain roast as he had to sell out, there were flaming gas jets at the front of the shop the butcher had a huge stripped apron and a straw boater. We always had a good Christmas, the butchers shops had rows of turkeys hanging along sides of beef'.

The butchers shop Brassells fondly remembered by Ena, was Percival Noah Brazil at number 172 the High Street on the east side and one of three branch shops. The shop was housed in a typical one story front extension added on to the original building, close examination of the photograph would suggest it was taken in early morning and possibly in mid summer. The presence of a Christmas club poster in the side window (not shown) is misleading as often they were in situ all year round. The shop blind and awnings have not been used on this occasion as maximum light was required by the photographer. The job designation of the three employees from left to right would appear to be frontsman, cutter and manager.

The latter named possibly a member of the Brazil family whose light protective apparel confirming warmer months. The employees' job titles were uniform throughout the London trade, although duties varied dependant on type and volume of trade.

A full complement of staff in the larger shops would consist of the following: manager, cashier, first and second frontsman, first and second cutter, driver cutter and improvers and any number of part time ancillary staff. The use of protective clothing dates from the middle ages at which time the wearing of aprons outside the shop was forbidden by order of the Butchers Guild. Shakespeare wrote "know you not being mechanised you ought not walk upon a labouring day without the sign of your profession".

The pattern of family functions within the Brazil business is difficult to determine but is believed that two sons of Percival Noah Brazil, Noah and Frederick ran the business of three shops in partnership. Noah age 59 and family lived in Lady Margaret Road and Frederick aged 40 lived with his widowed mother Susannah Brazil in Tufnell Park. Noah's son Albert H. Brazil managed the Queens Crescent branch aided by his sister Susannah with employee Albert Burgourne, a butcher's assistant, all of whom lived on the premises. The Queens Crescent branch was located on the southside in the market-end between Allcroft and Weedington roads. The Kentish Town road shop was sited on the eastern side three doors south of C. & A. Daniels drapers shop on the corner of Patshull Road. All three shops passed to new ownership soon after the First World War.

However amid the happiness and joy of inspiring festive shows it is was well to remember sometimes all may not be what it seems. Some unscrupulous proprietors would attend a well-advertised Christmas fat-stock show and purchase a prize-winning beast, ensuring maximum local publicity. A portion of meat from the carcass was then distributed to each shop with orders to mount a prominent display using the sample of meat as the centre-piece. In addition, each shop receiving a set of horns and rosettes kept specifically for this occasion. The customers were then charged, and were willing to pay a premium price for meat from this magnificent beast, or so they assumed. In truth the other beef in the display except the sample was of no higher quality or value than normally sold, and in this way, the deceitful trader not only recouped his initial purchase price, but also made excessive profits. To avoid suspicion, the genuine winning certificate was circulated between shops on a rota system.

The comment by shop staff 'the governor has bought a beast with six horns again', was a familiar cry in some businesses until prize-winning cattle reached such astronomical prices this particular deception became unprofitable. However this was quite a different issue to a genuine price increase in proportion to the wholesale cost incurred by the butcher for premium beef, lamb, pork and poultry at Christmas time. Fair trading butchers, who by far outnumbered the dishonest, always disclosed their intentions as in the Dewhurst advertisement where a rider informs the customer Christmas meat at Normal Prices.

One may question why employees acquiesced to such deceitful practices Quite simply they were coerced into believing by those in authority, be they shop manager or owner, these so called tricks of the trade paid for their free allocation of meat. As social reformer Charles Booth discovered 'butchers assistants although working long and irregular hours, they do not appear to have been poorly treated, receiving meat allowance to supplement their wages' He was obviously unaware how some employers recouped their losses. It should not be thought these malpractices, albeit uncommon were confined to any particular area or class of butcher, the perpetrators knew no boundaries except in their degree of sophistication and dishonesty.

G. F. KIMBER

Inside or outside, large displays of meat continued unabated through to the early 1970's. Most notable for volume until this time was George Frederick Kimber & Son 317-319 Kentish Town Road, who had the largest and most elaborate of these shows of any shop in the area. He acquired the business as a single unit in c1907, one of two shops housed in a large building until extended into the adjacent greengrocers in 1910. The busiest and largest open fronted shop of its kind in the road and in its heyday employing in excess of fifteen staff. Eye witness accounts from former customers and previous employees bear testimony to the high volumes of meat, poultry and game that were regularly displayed; a practice that continued after the shop was taken over by J. Ritchie. Former employee and local resident, Bob Boud, now retired after a successful career as owner of retail butchers shops and farm in Australia confirms, 'When I worked there about 1964-5 the shop also had a fruit and vegetable section with three staff, the meat side was very busy and had seven staff, plus part time Saturday staff. Out the back of the shop they made their own pies, sausages, dripping etc. It was terrible during the cold and windy months, the sawdust would get blown about because of the open front, and it got so bad at times that we would have to pull the shutters down'. The busyness of the shop confirmed by correspondent Ron Hardy who also recalls part of his duties involved preparing the numerous customer orders for home delivery. One among the sprinkling of celebrities he is unlikely to forget as he relates, 'While working at Kimbers I cut the meat for Yuri Gagarin when he was at the Russian Delegation in Parly Hill.' This most famous of astronauts, Gagarin (1934-1968) became the first human in space and the first to orbit the earth on the 12th April 1961; fêted as a national hero in his own country he toured several foreign capital cites where he was greeted with equal acclaim, including London in July of that year. The term 'Parly Hill' fondly used by Ron Hardy, the colloquial name loosely applied to the immediate area of streets and roads to the east of Parliament Fields in Kentish Town. Since his days working and living in the area he has achieved and enjoyed a successful meat trading business with his family in Toronto, Canada and now spends his retirement fishing and hunting.

The Kimber shop was also used to give young meat trade students practical butchery experience, as Mr Harry Memory of Walthamstow from an earlier generation remembers, writing in a trade journal having spent a day there in 1937 whilst studying for his Smithfield Institute diploma. The Institute opened July 1924 in Saffron Hill, gaining world renowned for technical education as the Smithfield College, and responsible for the establishment in July 1946 of the Institute of Meat. There is no evidence that George Kimber and Son were ever incorporated as a limited company, however he did form a partnership with his son Leonard with a shop at 278 High Road, Kilburn until, dissolved in January 1949 by mutual agreement. Interestingly their representative in this matter and all family affairs requiring expert advice were undertaken by their second eldest son

Philip Robert Kimber, solicitor to the meat traders' federation. The family lived locally for a number of years then comprising of George originally from Steventon, Berkshire and wife Amy that together with daughter Winifred and sons Leonard, Philip and William were all London born. And as was common practice until the First World War resident domestic servants Mildred Ball and Minnie Barnes. The business continued to trade until the early 1950's when the shop was sold, but not renamed, to butchers J. Ritchie. After almost fifty years service to Kentish Town residents, George retired to Worthing, Sussex, where he died aged eighty-five on 7th August 1960 Customer loyalty was such Kimbers name was kept above the shop until c1969 when sold by the Ritchie family. To this day, despite the long intervening years, older residents still refer to Kimbers old shop as a point of grid reference in the high street.

BUYING THE BEST

At the other end of the spectrum are the great meat displays in the food halls of premier stores such as Harrods, Selfridges, and John Barkers of Kensington, where once the Prince of Wales, later King Edward VIII, for a brief moment in history had his prize winning pig carcasses displayed at the Christmas show. The earliest exponent of multiple window selling is Whiteleys of Bayswater, the first of such stores to open in London in 1863. Barkers of Kensington and the Army and Navy store located in Westminster opened within a year of each other in 1870-71, later heading a group of department stores owned by the House of Fraser. Ensconced in fine buildings Debenhams, Harvey Nichols, Swan & Edgar, Hamleys toy shop and the swish Liberty store founded in 1875 by Arthur Lasenby that needs no introduction and where fabulous merchandise and stunning window displays, created by talented individuals from the like of Andy Worhol and Salvador Dali, represented the very best in British retailing. It was said of Harrods they could arrange a Wedding, Christening, or a Funeral and provide everything their customers are likely to eat, use or wear between birth and death.

Figure 87: Smithfield Club Cattle Show, Advertisement 1935

In an age where communication has reached untold heights of sophistication, it is perhaps understandable that we forget shop windows and country shows were almost the only means whereby not only butchers but farmers, could stimulate interest in their goods and animals. Of the agricultural shows and societies, which encouraged public and professional interest in food production, but by no means the only one is the Royal Smithfield Club founded in 1798. Their inaugural show for stock rearing and breeds began inauspiciously in a livery stable in Dolphin Yard, Smithfield, and from 1806 farming machinery of sorts was exhibited, after they moved to new premises in Goswell Street then a well known area for London's dairy farming. It is reputed showground proprietor Mr Sadler, of Sadlers Wells's fame, fed the livestock from proceeds taken

Figure 88: J. H. Dewhurst Ltd, 1971.

at the gate. This remained their home for the next thirty years until transferring to the Horse Repository near Portman Square. One of the founders of the Smithfield Club from 1825 until his death was Viscount Althorpe, John Charles Spencer, and third Earl Spencer (1782-1845) the 8th Earl Spencer; the late Princess Diana's grandfather.

From the beginning the show attracted a diversity of countryside and city dwellers, including agriculturists to retail butchers and machinery manufacturers, of local interest Mr Stanbridge Clark from Pancras Warf, Malden Lane, Kings Cross, hoping to increase sales of his wheat and potato manure. Foremost amongst the butchers were the Royal meat purveyors that in 1866 included Mr Collingwood of College Street, Camden Town, Mr Preston of Somers Town and later Kentish Town, who both purchased prise winning Sussex steers in their respective classes. At the same show prize winning Southdown sheep from the pens of Lord Walsingham were sold to the same Mr Collingwood for his branch shop in Upper Street, Islington, and similarly sheep to Mr Hooke of Frederick Street, Hampstead Road; and Mr Cridlan of St John's Wood. The following year buyers among the lower ranking retail butchers were William Cornish of Leighton Road, Kentish Town and Jabez Elvidge of Seymour Street, Euston, buying the first prize winners of short-horned steers in class 11 and Sussex steers and oxon in class 16 respectively.

Most notable in 1868 with regard to Kentish Town was master butcher Henry Read sited in the high street and other locations in London including the intriguingly named Liquorpond Street, Grey's Inn Road, Holborn. He purchased a first prize winning pen of nine and twelve month old white pigs together with a pen of sows exhibited on behalf of Her Majesty Queen Victoria. As farmers and land owners themselves at Windsor and Sandringham and always the object of Royal patronage, Prince Albert, dubbed a Smithfield man, was a regular and knowledgeable visitor and invariably had cattle on show from the Royal farms. Whilst visiting the 1840 show with his young bride Queen Victoria, one of the Royal cattle licked the hand of the Prince Consort as he paused to inspect the animals; that according to newspaper reports, so moved the Queen she purchased the animal from the butcher who had purchased the beast that very day and commanded the animal be spared and live the rest of its life in Windsor Great Park. Other sections of the press questioned the wisdom of such a Royal pet that stood 16 hands high of gigantic proportions and was anything but handsome.

At this stage the Smithfield Club annual Christmas shows were regularly attracting crowds of twenty-five thousand or more and disappointed prospective exhibiters were clamouring to show their livestock and machinery. It was obviously time to move on, and in 1862 a decision was taken by the Royal Smithfield Club to design and build its own exhibition centre. A two acre site was chosen in Islington where a new building, with a 125 foot span glass roof, was constructed, originally titled the Agricultural Hall, until the Royal prefix was conferred later. The last of more than seventy shows held at the Royal Agricultural Hall Islington was staged in 1938. Left to decay, it was rescued and refurbished in 1985, as the Business Design Centre. In October 1999, a plaque to commemorate the opening of the old Hall in 1862, was placed on the wall to the right of the entrance in Berners Road and was unveiled by two Smithfield Club members who had attended the very last pre-war show. After the war a new home was found at Earls Court Exhibition Centre, built in 1937, which they shared with Crufts Dog Show, the Motor show and other events.

In attendance at these shows were the titled land owners, in particular Sir Walter Gilbey (1831-1914) with brother Alfred in 1857 co-founding W. A. Gilbey Ltd, the company famous as producers of 'Gilbeys Gin' first marketed in eighteen ninety-five. The product became a household name and was produced at their main distillery and warehouse in Camden Town. Sir Walter and son Walter Henry (1859-1937) shared an intense enthusiasm for agriculture and used their considerable wealth in the pursuit of excellence in cattle and horse breeding. From benevolence towards institutions and charities, both men donated vast amounts of money to these and financed other deserving projects. In 1893 Sir Walter, who lived in some splendour in Cambridge House, Regents Park secured the freehold of 13 Hanover Square in London to ensure the Royal Agricultural Society had a permanent office. As Chairman of the Agricultural Hall Company, Islington he was instrumental in securing much needed space with the addition of a new hall, appropriately named the Gilbey Hall. In the history of English agriculture none shines brighter than the Gilbey name, regrettably all their achievements to extensive to record here.

Their business empire from shipping to jam making factories that financed their outside interests was by the nineteen-fifties struggling to survive and in 1963 International Distillers & Vintners was formed of a merger between United Wine Traders and Gilbeys Ltd. The inevitable reorganisation of the company spelt the end for operations in Camden Town, which after nearly a century was shut down and relocated to Harlow, Essex and all supply and distribution for retail outlets transferred to Wembley. In its time the Gilbey presence in Camden Town was of immense importance for local employment, the name survives in Gilbey's Yard, and the Roundhouse then rented from the railway. Finally in 1972 the Gilbey family relinquished ownership and the company became part of Grand Metropolitan Hotel Group.

Of interest from the political establishment the enigmatic Sir Oswald Mosley, future leader of the British Fascist Movement and regular show exhibiter of cattle and pigs reared on his estate at Rolleston Hall, Burton-on-Trent. In the 1904, show awarded fourth prize in class 88 for steers and King Edwards, Aberdeen Angus won the reserve. A political figure of notoriety in later years, his son Max Mosley at time of writing is president of the FIA, motor sport's governing body.

Over the years the participation at the main shows of the landed gentry and small independent retail butchers diminished. To be replaced by larger companies able to afford the increasing cost of show animals and maximise the potential publicity as for example; the London Co-operative Society in 1961 outbidding keen competition to buy the cattle champion 'Snowflake' for £1,300 pounds. A sign of ever changing times by 1980, when the supreme cattle champion 'Panda' was bred on behalf of Boots the Chemist. The supermarkets in particular never reluctant to exploit the marketing tool of advertisements emphasizing their pseudo bond with traditional retail

Figure 89: Turner 38 Fitzroy Road, Regent's Park, 1905.

butchers planted in the minds of the public. The supreme show champion in 1985, coincidently named 'Hae Presto' was purchased on behalf of Presto the supermarket chain fresh meat division. With the exception of newsworthy astronomical sums of money paid for the supreme champion in the cattle class, national newspaper coverage has gradually declined and is now mainly confined to local county newspapers and trade publications. For the most part the public memory, if not personal experience, is restricted to picture postcards illustrating the butchers shop window displays such as witnessed outside the Turner shop.

The road formally titled St Georges intersects with Chalcot Road, Regents Park where this particular shop occupied a corner location. Individual identification is problematic, however pictured with the possible exception of the butchers lad in the centre are three members of the Turner family who lived above the premises. The entire family in descending order of age are widower and proprietor William Robert Turner, eldest daughter Caroline, sons Frederick, Walter, Herbert and youngest daughter Florence. In addition the family employed the services of Battersea born Eliza Harding as resident cook and general domestic servant. A canny Yorkshire man, their father William had ensured the diverse career path of each son engaged as retail butcher, market meat salesman and butchers clerk covered the most important requisites for a successful business future. The surname Turner is associated with several butchers' shops in Camden dating from the 1860s.

Of minor interest the Whippet type dog in the bottom right hand corner of the photograph would indicate a family pet or a supremely confident stray. On the subject of pets and food premises, dogs and cats were once routinely kept by retail butchers and other food traders to suppress the vermin population. In slaughterhouses dogs were allowed under local bye-laws providing the ferocious type were muzzled and chained during the hours of business. The opportunity for meat scraps obvious to any canine, hence 'Fit as a butchers dog' a well known adage. However John C. Hotten, author of 'A Dictionary of Modern Slang, Cant and Vulgar Words', 1859, gives an alternative meaning to this accepted traditional saying "To be like a butcher's dog and lie by the meat without touching it; a smile often applicable to a married man."

As sales volume began to decrease, keeping the window and shop filled with meat each day became financially unviable. Instead row upon row of brown paper carrier bags with string handles began to fill the hanging rails, these brown shopping bags printed with the butchers stock in trade of meat joints and pastoral scenes of animals from whence they came. The blue ink prints a feint reminder when butchers shops dominated retail meat sales. The once favoured double windowed and corner sites became known as dust bowls and were now to be avoided

The butcher's silent salesman is still his most potent asset, but the days of ostentatious and some might say barbaric, displays of animal carcasses are virtually at an end and in all probability may soon be prohibited by law. The obverse of such changes has been to embrace the dictum 'small is beautiful' and butchers have responded with smaller, imaginative and colourful displays more pleasing to the eye; however one curious fact almost unnoticed by window shopping devotees, perversely the independent butcher's greatest trade adversary the supermarket rarely if ever have window displays. The customers must enter the store to see the goods on offer, just as our ancestors did centuries ago.

MEATHIST MINIATURE

The wild turkey fowl originated from Central and North America thence shipped to Spain aboard the merchant vessels. In the 16th century English trading ships with spices and other exotic cargoes from Turkey en-route from Spain bought these birds to England. They were first sold by London tradesmen dealing in exotic imports called Turkey merchants, hence the name turkey. Entrepreneur Sir William Strickland who is credited with creating the UK turkey industry, on receiving his knighthood incorporated a turkey on his coat of arms.

12

DELIVERIES DAILY

From very early on, customers if they so wished could have the convenience of a home delivery service provided free of charge by most retail butchers. The country town and village butcher in particular, where customers were widely dispersed and visits were less frequent, operated extensive delivery rounds throughout the outlying districts. In many instances these accounted for a major proportion of shop trade, although not all butchers provided this service. Those situated in densely populated street market areas with high volume cash sales and a pricing policy dictated by competition, rendered a regular delivery service financially unviable, but this was the exception rather than the rule, most butchers invariably displayed the sign 'Customers Waited on Daily', and did provide a daily or weekly delivery service.

Orders for customer requirements could be placed in person, by post, via the delivery driver, or by telephone as these became widely available. Terms were cash on delivery (c.o.d), or the option of a weekly or monthly account. With a reminder 'Prompt Payment Appreciated' for settling ones account printed on the invoice. To solicit new customers on the delivery round and update their regular home shoppers a list of prices, from hand bills to elaborate booklets, were distributed.

A considerable amount of one-up-man-ship was involved within certain sections of the delivery service; employers prided themselves on an efficient and speedy service believing smartly dressed staff and a well turned out conveyance ensured customer loyalty. Predominantly the task of men and boys, although in times of national emergency women often took over these duties; those employed on deliveries were as varied as the transport that was used. Some were time served butchers or apprentices, others were specifically recruited as roundsmen or butcher-drivers, alternatively untrained casual labour and in particular schoolboys. The latter confused with Errand or Delivery Boys the name given to those in full time employment.

Whatever their status the same golden rule applied for the wealthier customers, always use the tradesmen's entrance, a feature of the job that slowly disappeared as class distinction became blurred. Many roundsmen formed close bonds with their customers extending a helping hand to the infirm and elderly where needed. In the matter of settling customer accounts, many are the tales to tell from subterfuge to violence in extreme cases. Such a case involved the Hon. Theobald Fitzswalter Butler residing at Downshire Hill, Hampstead, who stood accused at Marylebone Court, in December 1842, of threatening to shoot James Wilbraham a solicitor's clerk. The clerk representing the solicitor engaged by Mr Southey, a butcher of Hampstead, for the recovery of monies owed for meat delivered to the defendant. In a trial of claim and counter claim the proceedings against Honourable Theobald were halted sine-die on his own securities unless any further charges be preferred against him.

The modes of transport used for home deliveries has varied through the ages from errand boy with wicker basket, the two man shoulder tray, hand barrows, bicycles, horse drawn transport, railways and motorised vehicles, including motorcycle combinations with specially adapted box sidecars; the ultimate in home delivery the mobile-shop that together with the hawking butcher can loosely be described as a delivery service. For the commonly accepted part of the delivery person's job, as a representative of his employer, was salesmanship and many carried extra goods with this in mind.

EQUINE ENERGY

The custom of carrying goods from house to house for sale is indelibly linked with the life and times of the itinerate trader, a way of life as financially precarious as any could be and given such names as peddler, tinker, haggler huckster and hawker. In essence, all travelling salesmen usually without premises and selling a variety of goods. Of the thousands that roamed the highways and byways some were unable to secure regular work, and for the majority the irresistible lure of being ones own master. For many falling on hard-times and treated as vagabonds and wastrels by sections of the public and authority and petitioning the Court for Relief of Insolvent Debtors, like Henry Hutcheson in May 1859 hawker of meat and occasional butcher of 20 Warren Street, 17 Upper Cleveland Street, John Street and other lodgings.

Figure 90: Meat Hawker

From among their ranks is the hawker, who dealt in meat that in the early days in London was almost entirely confined to a sales area within a comfortable walking distance from the markets in Leadenhall and Newgate. This type of selling developed into what came to be known as the 'Trotting Butchers', travelling into the surrounding pre-suburban countryside; the raw meat carried in a single basket positioned across the pommel of the saddle and secured with leather straps to the horse. The service was often erratic, operating only when the hawker was 'hung-over' with meat, a euphemism for being overstocked. The trade was gradually absorbed into the retail butchers shop, changing in character from one of speculative hawking to an order and delivery service.

For a while it assumed a flavour of the pony express in the days of the American Wild West, with men or young boy riders astride fast trotting steeds sent out to call on potential customers for orders and returning afterwards to deliver the meat.

This in turn stimulated competition between rival firms reflected in fierce haggling for the fastest and most reliable pony or horse at the market sales. In an effort to control the activities of the thousands of travelling salesmen selling all manner of items by the eighteen-eighties, the annual licence payment for a hawker with one horse had been increased to £4 pounds, 4 shillings.

A familiar sight until the late 1920's was the ubiquitous pony sized two wheeled cart, universally adapted for use by most trades, butchers, bakers, fishmongers, drapers among them. The driver's seat was reached by iron steps positioned by the curved shafts on either side of the vehicle, brass lamps were a prominent feature and the two small doors at the back of the body work gave access for the storage of meat. The catchment area for deliveries by Kentish Town butchers lay generally in the districts of Hampstead, Highgate and to a lesser extent Regents Park, the residences of the more affluent clientele accustomed to, and indeed expecting, this personal service. Within those areas the competition was formidable as exemplified in a photograph taken outside the premises of butcher and farmer Joseph Jupp at 158 Regents Park Road c1907, we see assembled five London butcher's light delivery carts. The proprietor Mr Jupp was in a substantial way of business with an extensive domestic, hotel and restaurant delivery service. For his high class trade, where reputation was based on quality and service rather than price, the prospective customers were given detailed information to assure them the retailer they had chosen dealt only in the choicest of meats and were businessmen of the highest integrity. A combined price list and illustrated booklet circulated by his roundsmen is an exceptionally fine example. Within the scope of a few pages he had managed to include breeds and origins and a guarantee of animal health, the majority of stock reared on the family farm at Godstone, Surrey.

Figure 91: J. Jupp, 158 Regent's Park Road, c1907.

Premier stores and prestigious retail butchers shops, especially royal warrant holders who counted among their clients the aristocracy, spared no expense when it came to matters of delivery vehicles using smart box-carts, driver and horses decked out in the finest liveries. Sound business reasons lay behind their apparent extravagance and the unknown author of the following statement leaves us in no doubt what they are:

> *"The butcher is a good judge of animals, he knows how to keep an animal in good condition, he chooses well fed animals for his customers, he is a judge of animals consequently his meat is good. A butchers who uses a decrepit pony and ramshackle outfit knows nothing of other animals, he must perforce be a bad judge of cattle and therefore not fitted to purvey meat".*

Accidents involving horse drawn vehicles were numerous and nearly always with serious consequences, so considerable horsemanship was required on the larger wagons. The carman, as the driver was then known, positioned towering high above the traffic controlling two or more horses, both driver and horse had little protection from the vagaries of the weather. A pleasant enough job in the summer months but on frosty winter mornings extra protection for man and beast against the more extreme elements could be little more than hessian sacks. The steep icy gradients of such roads as Haverstock and Highgate Hills and South End Road leading to East Heath Road, nicknamed donkey stand hill, presented a formidable challenge resulting in horses and drivers forsaking their meagre protection in the name of safety.

The sacking was placed beneath the wheels to improve grip, and eventually chisel shaped screws were inserted into horse shoes to prevent them slipping. The names of ancient inns and taverns sited at strategic locations a sign post to the formidable hills. The Bull & Last, Highgate West Hill and Cart & Horses, Haverstock Hill a staging post for man and horse to rest before the physical challenge ahead.

In 1929 Major General J. Vaughan circulated the concerns of the Institute of the Horse to a number of provincial and city newspapers including the St Pancras Gazette, imploring the relevant road authorities to make road surfaces less slippery. A forlorn hope, the institute meantime had successfully tested a special pad and shoe which they recommended as "suitable under all conditions"; how many horse owners took advantage of this offer is unknown. On medical matters concerning the horse Edward Coleman, principle of the Royal Veterinary College in its early days, devised a horse-shoe specifically designed to prevent diseases of the hoof especially contraction. The college situated in Royal College Street, hence the name, dedicated to the welfare of animals was often in need of financial assistance. From the many fund raising schemes, one from the imaginative mind of Sir Walter Henry Gilbey in 1935 stands out for its originality. It was named the 'Nose Bag' a farthing endowment fund, whereby forty or more veteran war horses with specially adapted army regulation nosebags, with their owners collected an estimated two-hundred and fifty million farthings for the fund.

Accidents due to human error were common and as today the majority attributed to a momentary lapse of concentration by either driver or pedestrian. One of many, the death of Mrs Frances Moore in September 1851, who died in University College Hospital after being struck by the butchers delivery cart driven by Thomas Perkins in the employ of Mr Dickens, master butcher of Camden Road Villas. The cart it was said driven at a furious rate by the youth although it was unclear if the accident was from misconduct by the accused or the horse running away and for this reason the youth was acquitted. The magistrate Mr Wakley observing the reckless way in which the butchers drove and rode their horses through the streets, as though empowered so to do by law.

The following from the police traffic returns for London in 1870 makes interesting reading: 124 persons run over and killed, 1,919 maimed or injured, a total of 2,043 and when subdivided by vehicle: 440 by cabs, 102 by omnibuses, 245 by broughams and carriages, 636 by light carts,

158 by heavy carts, 110 by wagons and drays, 257 by vans, 10 by fire-engines, 79 by horse ridden, and five by velocipedes, forerunner of the modern bicycle. The efforts of the Street Accidents and Dangerous Driving Prevention Society drew public and government attention to the intolerable traffic conditions in London; through society meetings where improvements and ideas were debated at length; not least the inadequate vehicle lighting regulations. At one such meeting the agenda included the danger to the public from fast-driving butchers' carts and railway vans.

An example of which in the case of Lewis Aggissn a young butcher charged with the manslaughter of Elizabeth Hedges by driving a butchers two wheeled cart furiously and knocking her down. The accident occurred in Highgate Road; the elderly women from nearby Chetwynd Road received fatal injuries. Similarly, butcher delivery cart driver twenty-two year old Walter Joiner aged 22 was charged with the death of six year old Arthur Elder in Leighton Road. At the time in February 1883 he was working for Henry Hook, master butcher at 317 Kentish Town Road and lived above the shop. The charge was eventually withdrawn following the inquest verdict of accidental death.

Figure 92: A. Pippett, Butchers Delivery Cart, c 1890.

From the 1860's onwards, as domestic trade expanded, delivery carts and wagons grew in number and variety and eventually refrigerated carts and vans. The early type like Acloms Patent refrigerator cart priced £35 pounds in 1877, using ice blocks contained in metal lined compartments incorporated into the bodywork.

Businesses including the meat trade soon realised the advertising potential and sign writers and commercial artists were kept busy adorning vehicles. The advertisements, which in some instances can be considered works of art, ranged from simply the butchers name and address to pastoral scenes of New Zealand sheep farming. Generally speaking mobile advertising remained conservative, manufacturing firms however became increasingly more adventurous, and in the best traditions of showmanship the delivery conveyance itself became the advertisement.

The visual impact of one, a team of oxen drawing a van advertising Hugon's shredded beef suet, travelled all over the country. Intended as a one off publicity stunt it so captured the attention of the public it was used until the 1950s. One shopkeeper became concerned for the welfare of the oxen until it was pointed out there were several teams of oxen and wagons operating in separate areas of the country. The product originally manufactured in Manchester, is better known under its trade name Atora and still widely available. There was of course an incentive to have your name and address emblazoned on your delivery vehicle, for those who did were exempt from the tax duty.

To service the needs of horse transport, a nationwide industry of crafts, trades and professions had emerged, veterinary surgeons, forage chandlers, wheelwrights, harness and saddle makers, blacksmiths, coach and wagon builders and so on. Within Kentish Town there were dozens of these trades: John Norris who lived over his harness shop adjacent to his trunk making business in Malden Road, and carried out work for the local butchers and other trades people. Surviving members of the Norris family, thought to originally hail from Wiltshire, still live in the area. The name Norris is one of distinction within the Worshipful Company of Curriers, several attaining the high office of Master from the 19th century onwards; the currier, or corriours, dresses and colours tanned leather.

The likes of Master farriers Henry Merralls & Sons at 63 Fortess Road and William Clark in Rochester Place would have shod the horses and donkeys, while other small workshops such as horse shoe makers William Cooke & Co. Ltd., Harbar Works, and horse nail makers Stanley S. Hardy Ltd., both situated in Little Green Street, Highgate Road supplied the means. Forage and bedding was sourced from the numerous hay and straw markets that existed in London at that time. The most famous the Haymarket in Charing Cross until moved to Munster Street, Regents Park. When first opened in 1830, known as Regents Park market, but officially titled the Cumberland Hay Market opening alternate days of the week. For busy local traders, orders could be placed at the London Corn and Forage Co offices at 268 Kentish Town Road.

In the classified section of newspapers could be found advertisements for horses wanted or for sale. In a wanted advert placed by wealthy butcher Mr Pippett in 1878; "A pair of strong and fast well matched roan or dark brown horses, preferably quite and friendly, sound and not over six years old". The wording indicating the animals were required for carriage driving and not for delivery rounds.

Figure 93: Page's Delivery Invoice, (section) c1930.

An indication of the numbers used throughout the retail meat trade in London at this period, and of special reference to Camden, can be judged by those kept with associated horse accessories by the following retail butchers: The O'Hara family had 7 horses, 6 carts, 1 horse van for their shops in Leighton Road and Highgate Road and 8 horses, 6 light carts for their shops in High Street, Hampstead and Kilburn. George Grantham with five outlets in Park Street and High Street, Camden Town had 7 horses, 7 ponies, 5 carts and 1 horse van while his near neighbour Edward Wright in Park Street, had 6 horses and 6 carts. Lastly James Collingwood with a double fronted shop in York Road kept 3 horses together with carts stabled in the mews at the rear of 3 Clifton Road, Camden Town. In addition all the fore mentioned and common with most successful businessmen had conveyances for private use.

As again with master butcher Matthew Knight, at 206-208 Kentish Town Road, the owner of a wagonette, that on one occasion had side lamps stolen by a dishonest employee. The culprit Nathaniel Colson left his employ without notice taking the property which he subsequently pawned for 5 shillings at Thompson's the pawnbrokers in Chalk Farm Road. The carriage lamps were recovered, leading to the conviction of Colson at the Middlesex Assizes on a charge of theft. At his short trial in September 1877, Mr Fletcher the magistrate noted his previous criminal record and sentenced him to eight months hard labour. The wagonette and horses were stabled in Wosley Mews, a narrow cobbled lane which still exists behind and parallel to the high street between Gaisford and Islip Streets. Charles Knight, the founders' grandson and owner of stables numbered 1, 2, 3, 4 and 6 in the Mews, rented them to local businesses including nearby trade adversary, butcher John Edwards. Rather macabre numbers 1 and 3 were former slaughterhouse sites; nevertheless it was common practice to site both slaughterhouse and stable in close proximity to one another. The wagonette was a four-wheeled passenger vehicle of various sizes drawn by a single horse or pair and popular as a family carriage from the early 1840s until the end of that century.

From earlier on in 1788, Kentish Town was Mr Hopkins a noted London horse dealer trading from Spring Place with a substantial business in Red Lion Yard, Holborn. After his death the Spring Place estate was listed as follows: Newly erected stabling for 44 horses, dry feeding stall for horses with fever, excellent dwelling house over the stables with a number of rooms for servants, A farriers' shop, a large garden stocked with the choicest fruit trees, a spacious paved yard in which is a well, with curious engine pump that conveys the water from a reservoir to the premises. Also a new erected dwelling house opposite comprising three bedrooms, two parlours, a kitchen, wash house, a good garden with fruit trees, a summer house, with good cellar under the same.

The bulk of horses were acquired through horse fairs or markets and sold, resold or exchanged for being too big, too small, too fractious, too docile, too old, too young, too anything, but suitable for the owner. Horse bazaars, as places of sale were then called, existed all over London; one in Grays Inn Road in particular was well attended. Officially titled the North London Horse Repository, the building served many uses in its time, a co-operative bazaar, assembly rooms, brewery warehouse and even hosting an exhibition of Madame Tussaud's waxworks.

The equine population reached its zenith in 1902, with an estimated total of 3 ½ million horses in England alone, engaged in farming, hunting, delivery services, public transport and the army. The greatest number of horses was not attributed to personal and public transport as one might expect, but to the distributive supply sector and shop delivery services.

A happy state of business affairs for Thomas Roberts at 6 Spring Place, Kentish Town, who advertised his, services as heavy or light butchers horse drawn carriers. This service was primarily used by those butchers who had neither the financial means nor space to accommodate their own transport, instead using a contractor to collect their supplies from Smithfield market. For those with only one horse, versatility was the keynote, the same animal being interchanged between different sized vehicles.

A seemingly minor, but in actuality important, service that was provided at all London wholesale markets, was by the cart minders and whip-holders. To whom buyers would entrust their property for safekeeping in exchange for a small payment. The reason illustrated, and by the far from an isolated incident, was that experienced by Jabez Elvidge when Bejamin Harris was charged and found guilty of stealing his horse, cart and fifty stone of beef. The cart minder Charles Wright hired by Elvidge caught the defendant in the early hours of the morning on 1st January 1869 in the act of driving the vehicle away. A persistent offender, Wright was sentenced to five years' penal servitude. Mr Elvidge opened his retail meat business late in the 1850s, trading as Jabez Elvidge from premises at 12 Upper Seymour Street, then a continuation of Eversholt Street until a subtle change of name spelling and possibly road numbering finds Jacob at 102-104 Seymour Street, which he shared with wife Jane, children Albert and Jennie together with two journeymen butchers, a bookkeeper and two domestic staff. It was apparent early on the quality of meat sold would be of the highest standard bourn out by frequent buying forays to the Smithfield Club Show, in one instance the purchase from the Duke of Sutherland's of his first prize short-horned steer. There were two other Elvidge butchery concerns, one already mentioned in the Kentish Town area the other in Caledonian Road, but in the absence of evidence to the contrary they must be treated as separate identities. The big haulage contractors like the Union Cartage and Edwards provided for the majority of independent, multiple and chain butchers. The last freelance butchers carrier firm in Kentish Town, is believed to be Mc Kays Transport c1969 and based in Brecknock Road.

ELVIDGE & SON,
Family Butchers,
SEYMOUR STREET,
PRIME ENGLISH BEEF AND MUTTON
(Our own Breeding and Killing.)

Figure 94: Elvidge & Son, 1890.

With the introduction of massed produced vehicles powered by the internal combustion engine before the First World War, a gradual decline of horse power in favour of motor vehicles began, accelerated by a substantial loss of horses during the conflict. The petrol driven vehicle began to outnumber horse drawn cartage transport, although another thirty years or more would elapse before horses were completely replaced by their mechanical equivalent. Whether drapers C & A Daniels in Kentish Town Road had already started to convert to petrol driven vans when they advertised 'Light Two Wheeled Carts' in good condition for sale, in the Camden & Kentish Town Gazette in 1912 is not clear. The fact that more than one was on offer and all still serviceable seems to suggest they had accepted the inevitable. However in the same newspaper C W. French & Sons, Jobmasters at 65 Camden Road, had decided to err on side of caution when they advertised the following; "Motor Carriages and Horse & Carriages to let by the hour, day or year".

The potency of horse conveyance demonstrated by the detailed regulations to be observed for Hackney Carriages Standings within the Metropolitan Police district, from which a brief example has been extracted : (1) Kentish Town: Bartholomew Road, Three vehicles by dead wall, the head of the first horse where wall rises 20 feet. (2) Malden Road, Four hackney carriages extending southwards, horses heads facing Queens Crescent. (3) Grey's Inn Road. Twelve vehicles in two portions front of vehicles facing east, first portion two cabs on east side of public convenience commencing opposite No 4 along the kerb on south side of Theobalds Road. Second portion, ten cabs along the kerb on the south side of Theobalds Road, west of the public convenience.

Throughout the equine world are a multitude of societies and association either dedicated to maintain standards of breed, driving skills or purely social. Most of them established in the 19th century, the formation of many in which the first Sir Walter Gilbey, took a prominent part; The Shire Horse Society, Hackney Horse Society, Hunters' Improvement Society and of wider interest and enjoyment to Londoners in 1880 the Whit Monday Cart Horse Parade. Later he co-founded the London Harness Horse Parade, and became a governor and Principle of the Royal Veterinary

College. Using his unrivalled knowledge of horses he wrote many books on the subject "The Old English War Horse" pub 1888, required reading for horse enthusiast and historians alike. With his son Walter Henry, who subsequently held several identical society and agricultural Chairmanships, they proved a formidable combination of influence. In 1904 began the London Van Horse Society Parade held annually on Easter Monday in the inner circle of Regents Park. Although numbers of entries were restricted in earlier years a review of the turnout is not without interest as a barometer of changing trends. Entry from the stables of London tradesmen numbered 132 drivers at the inaugural parade increasing to 1,058 in 1914. By 1927 entries were down to the 621, of these 521 single horse vans, 85 pair-horse vans, and 15 barge horses, shown in working harness. Among the 113 general carrier vehicles were 67 bakers, 65 Greengrocers, 40 dairymen, 29 laundries, and 12 butchers, the brewers and wine merchants accounting for 30 teams and vans, provisions stores 29, and railway companies 49. In 1950, entries in general had drastically declined and only two butchers and fishmongers paraded. The event initially for trade vehicles gradually took on a carnival atmosphere with all manner of horse drawn vehicles attracting thousands of spectators. However by the mid nineteen-sixties, show entries together with the Cart Horse Parade began to dwindle leaving no alternative but to amalgamate. From 2006 the two events under the name of the London Harness parade were moved outside London to the South of England Centre, Sussex. Tragically the horse versus motor received an unwelcome boost during the First World War when government transport agents wandered the streets with discretionary powers to commandeer horses.

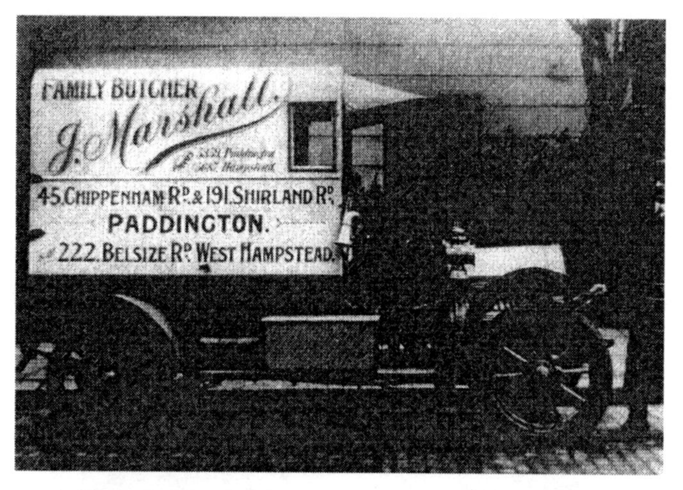

Figure 95: J. Marshall, Delivery Van, c1910.

Just how long the change over to motorised transport would take no one really knew. In the meantime there was still plenty of work for both modes of transport as the advertisement on behalf of cart, van, and motor body builders William Parkyn & Sons in Holmes Road testified. Nostalgia is a seductive mistress and there was of course the problem of pollution, noise, traffic accidents and congestion associated with horses, similar to those encountered from motor vehicles. Albeit crossing-sweepers employed to clear a path across the roads through the horse dung, was not one of them. A problem compounded when the Metropolitan Cattle market open in Islington. The market improvement committee advertised for Scavengers and others willing to take away dung and soil from the market-place, and to sweep and clean the roads and footpaths therein and surrounding areas.

Nevertheless Camden Town trader, John Turnbull of 185 High Street, sent an irate letter to his local paper which appeared under the heading "Vibration In Shops", complaining of 'shaky windows and his displays therein' caused by motor buses and heavy traffic'. He obviously had experienced both types of transport, and by the tone of his letter he preferred the horse and cart.

Figure 96: Delivery Van, G. Runnicles 65 Osnaburgh Street, c1950.

The reluctance of some individual tradesmen and companies to change was not always from any feelings of nostalgia, but in a genuine belief that horse transport would never entirely disappear and in any event the two could happily co-exist. The St Pancras council refuse collection service depot, at Suffolk Wharf stables in Jamestown Road, retained their remaining four heavy horses until October 1950, before selling them to a farmer for £165 pounds; the three stablemen who looked after them were kept on the payroll and offered alternative employment. One of the fleet of council dustcarts, painted green and instantly recognisable by the semicircular or D-shaped body with sliding shutters, could be seen positioned horse-less at the eastern end of Queens Crescent weekly market, the horse returned to the shafts at the end of each day's trade.

The horses used for such work by local councils were previously known as vestry horses, so named because financial provisions for their upkeep were administered by the local vestry. However in the St Pancras vestry minutes for 16th July 1876, the Metropolitan Drinking Fountain and Cattle Trough Association asked the vestry to contribute £6 pounds per annum for each fountain, which they politely referred the matter to the Highways Committee. A responsibility passed on when St Pancras was made a Borough in 1900, and with it their predecessors unjustified reticence to squander rate payers taxes when just over a quarter of a century on it was reported, councillors nearly came to blows over who should be awarded the contract for shoeing horses.

Figure 97: Cooperative, Mobile Butchers Shop, 1923.

The advances made by motor manufacturers was interrupted by a short revival in horse power due to petrol rationing and associated restrictions during the second world war which at the beginning, according to Ministry of Transport figures, the horse population of Greater London was still 40,000. How many of these were originally kept purely for pleasure is unknown, but from necessity most were quickly found useful employment. This was a great relief to their owners, as only working horses received official rations, amounts allocated depended on work undertaken and the size of the animal.

In contrast to the First World War when army requisitioning authorities searched the streets looking for mounts, although casualty rates of civilian and army horses in the second conflict including mules was relatively low. On the home front, their use made a valuable contribution to the war effort in conserving petrol and oil stocks, be it drawing post office vans in the city and west end of London or pulling the local milk cart. The peculiarities of war however saw bemused onlookers wondering why horses were positioned between the shafts facing the wrong way. This officially recommended procedure had a serious intent, to prevent the horse bolting during air raids. There is no evidence discovered so far, of any horse delivery service being used by retail butchers in Kentish Town, either during hostilities or post war.

A return to peace time would see the virtual end of horse transport as thousands of young men and women in the regular and auxiliary services had been taught how to drive motorised vehicles, and we may assume the convenience of a starting handle preferable to feeding, harnessing and mucking out stables at unsociable hours. It is noticeable in rural districts post war many village butchers still considered motor vehicles unnecessary in country districts where the pace of life was slower. In outlying districts there had existed from early times the 'cutting cart' that performed a perfunctory service as mobile shop These gradually evolved into an assortment of Heath Robinson contraptions, usually of box type single horse drawn vans. The first mobile butchers shop in the modern context, to be specifically designed for the meat trade, was constructed by the Co-Operative Wholesale Society and shown at the Edinburgh Congress Exhibition of 1923. The range and type became ever more sophisticated with all round windows, awnings, counters and refrigeration, in fact a complete shop on wheels. Under certain trading conditions the advantages of taking the shop to the customer had not escaped our own technological age; the new Dewhurst Ltd meat company est. 1995 had several mobile shops operating in different parts of the country.

With the exception of items preserved in museums, the remaining paraphernalia associated with horse transport has disappeared from sight. The horse will always remain a much loved and respected animal and will always be used for a whole host of leisure activities; their ancestors were truly the work horses of the nation. A long time past, when it was the custom to record distances from well known points, the Mother Shipton, public house in Malden Road was 2 miles 3 furlongs from Oxford Street, universally renowned today as one of life's great shopping experiences. In the post-war period petrol vans and lorries replaced horses for retail meat deliveries, popular with butchers the new Austin Ten van costing £275 with £7.10s extra for a standard colour of your choice.

Hardly envisaged in those times was the full impact of motorised traffic congestion, which at times for later generations threatened their livelihood; from the imposition of customer parking restrictions and in extreme cases to the actual demolition of the building. One angry butcher Stanley Kite, and his wife Dorothy, protested against a proposal by the recreation and open spaces sub-committee of Camden to demolish his shop. The removal of the building, on the corner of Castlehaven and Chalk Farm Road, required as part of plan to re-route local traffic. This particular shop also traded under the name Leonard P. Hillmam.

DISTANCE NO OBJECT

The potential for railways to transport livestock, and later on carcase meat, was quickly realised and began another episode in the history of the meat trade; with cattle becoming a familiar sight at the Midland Railway coal depot and cattle docks in Kentish Town Road; the intrusion of this particular railway into the district scaring the urban landscape for ever and only subsiding with the acquisition of the last remaining acres of farmland. The English railway began tentatively between Stockton and Darlington on 27th September 1825, using open carriages. Considered a madcap undertaking even for passengers, to convey livestock in such a way was deemed impractical. The doubters quickly realised the economic possibilities and railway transport of livestock began in 1831, within a short space of ten years the Liverpool and Manchester lines alone were handling 100,000 animals annually. The Journal of the Royal Agricultural Society in 1858 had this to say 'Fifteen years ago there were no railway communications between Norfolk and London. Cattle and sheep for the Smithfield Monday market had to leave their homes on the previous Wednesday or Thursday week. Such a long drift, particularly in hot weather, caused a great waste of meat. The heavy, stall-fed cattle of East Norfolk suffered severely'.

The average loss on such bullocks was considered to be 4 stones of 14 lbs, 'while the best yearling sheep are proved to have lost 6lbs of mutton and 4lbs of tallow; but beasts from the open yards and old sheep, with careful drovers, did not waste in a like manner'. Stock now left on the Saturday and were in the salesman's pens that evening. The cost of rail was considerably more than the old droving charges; but against the gain of 20 shillings a head on every bullock a Norfolk farmer sent to town, saying nothing of being able to take immediate advantage of a dear market. By the mid-1920s the number of animals transported by rail assumed the colossal proportions of approximately 3,000,000 cattle and 10,000,000 sheep.

Figure 98: Open and Closed Livestock Wagons, Liverpool & Manchester Railways, 1833.

From the beginning, a long running battle ensued between the meat industry and the railways companies over rates charged, supervision, welfare and un-hygienic condition of wagons used for transportation; the movement and volume of meat and meat products, another facet of meat delivery. The midlands based Palethorpe Ltd using G. W. and M. L. in 1937 negotiating freight charges with the Railway Clearing House, London; for per-ton of sausages, tongues, brawn, hams, pressed meat, pies and black pudding. In this case for a van part of a dedicated fleet belonging to the company emblazoned with their name and a packet of sausages painted on either side. The Palethorpe railway vans continued in service until the nineteen-sixties.

The parcel service adding an extra dimension to the business of butcher Henry Hooke and Sons 'We are close to all the Great Railway Termini. Joints of meat carefully packed for sending or carrying by rail'. The great termini, the customer convenience of which Hooke painstakingly pointed out began with Euston station opening in 1837, followed by Kings Cross in 1852, and St Pancras in 1868.

Theft was a particular problem for the poultry trade who protested constantly about mishandling and lost cases and hampers. Extracts from a goods managers report on the subject at Euston terminus in 1853 confirms that stealing was rife. 'Thieves are pilfering the goods from wagons to an impudent extent. Not a night passes without wine hampers, silk parcels, draper's boxes or other provisions being robbed'. To this catalogue of every day thievery can be added the prospect of railway trucks lost in transit, where consignees were left in the ludicrous situation of advertising in the personal columns of national and local newspapers. The appeals were generally to sidings owner to check the serial numbers of any trucks in their vicinity.

The excessive rates charged and injury to animals amounted to thousands of pounds annually, consignment charges in general across the whole system were a matter of the most serious concern. Loss or injury of animals caused through negligence was limited to £15 pound per head of cattle, £2 for sheep or pig and £10 for a horse, unless a higher value was agreed between parties beforehand. An obvious problem here was proving negligence, fractious animals or bad loading as it is self evident, accusations could not be corroborated. The matter of using quicklime for disinfecting their railway trucks, for which they charged one shilling a truck, and their refusal to use neither straw or sawdust or anything else over the lime made cattle and sheep liable to injury from slipping on the floors, and with pigs particularly receiving burns while lying down. Progressive governments and endless Royal commissions were only marginally successful in appeasing what were after all problems between two major industries, albeit of vital importance to the economic prosperity of the country. An indication of the difficulties is illustrated for example with regard to rate charges.

Figure 99: H. Hooke & Sons, 1914.

The production of home produced livestock was a fragmented business, thousands of farmer's countrywide sending small numbers of animals by rail. Imported cattle arrived at the ports of entry on mass and were transported to their destinations in bulk lots; wholesale importers therefore enjoyed cheaper rates on their consignments. Basic rates charged by individual rail companies were calculated on distance travelled, number of animals and extra for fodder and labour were also at variance. Singled out was the London and North-Western railway company, of whom it was said persistently charged more per wagon on their service from Liverpool than any other company. These and other complaints overlooked the fact that by and large the railways provided a speedy and efficient service. The old open livestock wagon exposing animals to bad weather gradually progressed to specially designed partly enclosed trucks and double-deck units with spring buffers replacing solid buffers, minimising the risk of jolting and subsequent injury. A further service to customers, and importantly a cost effective measure by railways in 1873, was the introduction of standardized trucks, available in three sizes small, medium and large with a recommended load capacity for each, dependent on animal size and species. Aristocrats of the cattle world, prize show animals, were given five star travel accommodation aboard superior type railway wagons; introduced as Passenger Train Cattle Vans from the 1880s, as the name implies coupled to passenger trains for stockmen and owners to accompany them. By 1922 there were eighty different types of cattle wagons. The livestock wagon continued to be built by British Rail with 3,800 new wagons bought into service between 1949 and 1953 when construction abruptly ended as cattle traffic rapidly declined. In 1972 the few remaining were withdrawn from service.

Incorporated in the building plans for the new Smithfield market, was an underground goods station provided by the Great Western Railway alongside the Metropolitan Line, the world's first underground railway opened in 1863. The Smithfield venture linked the market with other main line railway stations. An ingenious though complicated system of steam operated winches and hydraulic lifting mechanisms were employed to raise the meat from the basement station level sidings to the sales floor of the market. However deliveries by road transport from other railway companies continued to increase direct to the butchers' stalls. A futuristic idea for its time, the goods station or Rotunda its official name, never fully realised its early promise and slowly fell into disuse, attempts to revive its fortunes after the Second World War failed and in 1970 the Rotunda was converted into an underground car park.

PEDAL POWER

Pedal power was not forgotten and became the mainstay for local deliveries. A rare glimpse of two boxed tricycles outside the premises of A. J. Morley, Marchmont Street shows considerable dexterity was needed to use these vehicles when fully laden. In the photograph of Fortess Road, Kentish Town, the encroachment of motorised vehicles is well underway. Whilst in the foreground two bicycle roundsmen outside the shop of Ernest Vincett are preparing to leave with the morning's meat deliveries. The shop numbered 116a, sited on the east side of this road, was previously occupied for many years by butcher Thomas Budd Musgrove and wife who proudly announced in October 1898 via the Times Newspaper the birth of a daughter. He was one of several Musgrove's in business in this road, and further afield including cheesemonger Henry Budd Musgrove, recorded much earlier in 1874, trading from premises at 42 Hornsey Road, Holloway.

From the 1890s bicycles were increasingly used and by the 1920's virtually every butchers shop had at least one trade bicycle with a willow basket. Other models, all without gears, included the capacity for larger basket loads by reducing the circumference of the front wheel and fitted with a cumbersome support stand. These proved unpopular with delivery boys who preferred the standard wheel size, a panel incorporated within the frame displayed the owner's business address. The carefree attitude of the butcher boy on his round often hid the underlying fact they were lawfully employed and subject to all the legalities this entailed. The Weights and Measures inspectors made frequent on the spot checks whatever the mode of transport, a well understood legal procedure to ensure compliance with the regulations.

Figure 100: A. J. Morley, 36 Marchmont Street, 1903.

However total ignorance surrounded the employee's legal responsibilities with regard to private use of trade bikes. An innocent benefit which attracted young men and particularly boys to the job and that in 1932 incurred the intervention of several high court judges. In that year the casual attitude of some employees was given a jolt when a prestigious multiple retail butchers company trading in London faced a claim for damages. The plaintiff claimed for personal injuries caused by the negligence of a butcher's delivery boy who at the time was using his employer's trade bike during his lunch break.

It was stated the shop manager had given his permission and at an earlier hearing the injured party was awarded £300 pounds. On appeal the law Lords ruled 'When an employee, for his own purposes, used his employers' bicycle, by the employers' permission, the employers were not liable for the employee's negligence'. In explanation of their decision, under the Shops Acts the employer had no power to say how the employee used his dinner hour unless he was doing something in the course of the employers' business at the time.

Figure 101: Butchers Trade Bikes, Fortess Road, Kentish Town, 1910.

Always a highly competitive trading area, food outlets in Fortess Road, faced strong opposition commencing with Henry Charles Rouch, owner of several shops in this road trading as provision merchant, pork butcher, fishmonger, greengrocer and cheesemonger, who for a short time was in partnership with relative Mr A. Rouch, (forename unknown) until dissolved in November 1888. At this period in the eighteen-hundreds a successful business man living above the cheesemongers premises with his wife and daughter, and renting rooms to four domestic servants. Adjacent to the Roach home was butcher and proprietor Robert Wilson then in his sixty-eighth year and originally from York, living above the shop with his Suffolk born wife Hannah and son Frank and lodgers Lawrence Forrest and Walter Scholefield, all trained butchers working in the shop.

Between this run of shops which included grocer John R. Herbert and Willis & Son corn dealers lay Gottfried Mews the workplace of Arthur Fearn, coach builder. The mews takes its name from the proprietor of a bakers shop nearby Phillip Gottfried & Sons. The 1901 census reveals his age as seventy-four and his wife Sarah eighty-two, both living above the shop with other member of the family; German born Phillip was a naturalised British subject by the time of this census. On the west side lower down at the south end and of no direct competition were butchers shops run by Albert Gayes and Henry Hurst.

A mention of David Greig Ltd in the same road, est. 1870, although principally trading as tea and provision merchants also operated a chain of butchers shops. This side of the business considerably enhanced in 1962 on amalgamation with Colebrook & Co, butcher and fishmongers. In 1974 the Greig business, through a complicated series of take-overs, eventually became part of the Fitch Lovell, food industry group. Within this group eventually twinned with Keymarkets as David Greig Supermarkets until finally being absorbed into Keymarkets and loosing their trade identity.

From a similar business background the grocery firm of Williams Brothers, also trading as William Brothers (Butchers) Ltd, a private company with considerable plans for the expansion of their retail meat sector during the nineteen–sixties that succumbed to the financial overtures of Sheppey Trust subsidiaries together with their sister companies Williams Brothers' Direct Supply Stores Ltd. A decade later and another buyout, terms were agreed and food wholesaler Booker McConnell acquired the butchery business. In January 1980 meat wholesaler and importer Thomas Borthwick & Son Ltd. announced the acquisition from McConnell of the Williams Brothers butchery chain. Their fifty-eight retail high street shops merging with Matthews Butchers Ltd, a subsidiary of Borthwicks. In the same month, almost as an adjunct to the main course Borthwicks also swallowed up the seventeen butchers' shops owned by James Blue Ltd and Kelday Butchers Ltd, trading mainly in North West London; all of whom are no longer in trade.

MEATHIST MINIATURE

The origin of the Smithfield Stone, once used by the market wholesalers derives from calculating the imperial stone of 14lbs of live weight of a butchers beast is equal to approximately '8lbs of dead weight', the Smithfield Stone. This ancient method of weighing meat reputed to date from the 17th century became illegal on the 29th October 1939, and for sheer volume Smithfield market, before refurbishment in 1991, reigned supreme with 15 miles of hanging rails capable of displaying 60,000 sides of beef.

13

A GOOD NAME

A common feature among retail meat traders was their involvement with stock farming and like many of their contemporaries, the Coggan family made the transition from being purely a producer of meat to ownership of retail butchers shops towards the latter end of the nineteenth century. The family ancestry however pre-dates this period by several centuries, but our interest begins with George Coggan 1795-1870, farmer of Curry Rivel in the County of Somerset. In a small way of farming, mostly arable with just a few head of stock, any thoughts of dealing beyond the county boundaries and in particular the London markets was deemed unviable; as depending on route and stock collected it took the drovers ten days or more to reach their destination, for example travelling via Salisbury, Winchester, Basingstoke, Hounslow, and finally along the Marylebone Road to Smithfield. It had been calculated the average weight loss per animal was 20 lb for every 100 miles travelled, which would considerably reduce a farmers profit, despite attempts to redress the balance with copious amounts of water on arrival.

The change began for the Coggans with the construction in 1827 of the Bridgwater and Taunton canal, enabling barges to transport huge cargos of coal, timber and other commodities in turn attracting workers and businesses to the region. It was during this period that George became actively involved in cattle dealing and supplying the local butchers' trade. His business greatly improved as the network of railways spread further and further becoming the work horse of the industrial revolution. The canal-borne traffic in goods decreasing in consequence blighting what had hither too been prosperous farming and village communities. The country markets overflowing with animal stock produce and all manner of merchandise, a great deal locally made was either sold at knockdown prices or returned to the owner; the railway leaving in its wake unemployment and taking the men and women to the towns and cities in search of work. Among them landowners and businessmen seeking new markets; like George Coggan, together with his son Frederick who began to transport cattle by rail to London, the eldest of eight children Frederick James Coggan, in partnership with his father George, expanding the meat trading side with London. In eighteen forty four Frederick married Anne Gristock and started a family, a son William was born on the 19th July 1851 at the family home in Creech St Michael.

CONSUMMATE TRADER

The youngest of their four children, William Coggan would eventually prove to be the consummate meat trader, controlling every aspect of the family business from farm to final consumer. This

Figure 102: William Coggan, O.B.E.

wealth of experience, coupled with his natural abilities enabled him to become an extremely successful business man, representing the industry at the highest level on every conceivable subject concerning the trade. Like so many that had gone before, the road leading from Somerset to London was taken purely for commercial reasons and at his fathers request; William was required to be permanently based in London and oversee the family business interests on Smithfield market.

The passage of time has clouded the exact sequence of events, but it is believed the opportunity arose to extend their operations into the retail sector when payments for a substantial amount of carcass meat supplied to a retail butcher became owing to the Coggan family. The name of the insolvent butcher and precise details of discussions to negotiate payment are unknown, however it is reputed that agreement was reached to accept a shop in lieu of payment. The settlement of debt in this manner was far from unusual as inept retail butchers, faced with a refusal of further credit at markets would simply sell their business on the spot, provisional agreement often sealed with a handshake. Nevertheless the validity or otherwise of this story the research facts reveal William is recorded as ratepayer/occupier and retail butcher in St Pancras, the property formerly in the occupation of Elizabeth Cotterall.

Thus William Coggan aged twenty-three and newly married to farmer's daughter Elizabeth Clothier, made his first foray into the retail sector and in October 1873 we find him occupying 27 Delancey Street, Camden Town, trading as W. Coggan family butcher. The rooms above the shop provided accommodation for the young couple where Elizabeth gave birth to two of seven sons, the second eldest Cornish Arthur born in 1876 the father of former prelate the late Donald Coggan, Archbishop of Canterbury. The premises also housed shop and domestic staff, the oldest fifty-five year old general servant Catherine Skrine from Devon, and the youngest fourteen year old butcher boy Walter Frowed, a local born lad. In addition Coggan employed four butchers who lived away from the shop.

It is quite clear, even at this early stage in Williams's career, events were moving rapidly, because aside from his cattle and retail business responsibilities, local politics beckoned and in June 1879 he was elected vestryman representing West St. Pancras Ward. A political platform he shared with future Kentish Town department store owner Herbert Beddle, elected on the same day. His initial success however was short lived and at the end of his inaugural year William did not seek re-election.

In 1882 he sold the Delancey Street shop to butcher, John Fowles, having acquired partnership interests with Harris Crimp in two other shops at 4 Motcomb Street trading under the style of Harris Lidstone founded in the 1830s and 5 Pont Street both in the affluent district of Belgrave

Square, the Pont Street shop became his home address trading as W. Coggan & Co. In June 1885 the partnership between Coggan and Crimp was dissolved and William assumed full ownership of the respective businesses. A seemingly inexhaustible appetite for business acquisitions William is again noted in July 1886 on the dissolution of another partnership with Abraham Oscar Tyrell carrying on business as Coggan & Tyrell at 10 Woodstock Road, W.1, followed soon after with the acquisition of 19 Sussex Street, Warwick Square.

At this junction we may pause to ponder had William embarked upon a deliberate policy for future ambitions yet to be realised or simply reacted to a combination of random events. In hindsight we know this group of shops under the direction of William Coggan eventually became the nucleus for Lidstone Butchers Ltd. And it is certainly true the constraints of local domestic political affairs were discarded for issues of national importance. Be that as it may, a man of enormous ability, William was certainly never content to rest on his laurels. The following illustrates just how involved William, a freeman of the Worshipful Company of Butchers since January 3rd 1884, would become in shaping future events in the meat trade.

In 1888, on the 12th March, a meeting was convened at the Wellington Hotel, Dewsbury, Yorkshire by prominent members of the meat trade, although it is unknown if he was in attendance. Among the various items on the agenda included a resolution for the desirability of forming a National Federation, proposed by Mr W. Field of Dublin. In addition, it was also proposed that the trade should have its own trade paper and both of these proposals were adopted. Subsequently, the first edition of the Meat Trade Journal & Cattle Salesman's Gazette appeared on 5th May 1888, as a weekly publication. Less than three years later, in 1891, bowing to pressure from London trade figures Coggan amongst them, it was decided to move the whole operation to Smithfield, London; also agreed at this time the trade journal should be owned by people directly involved in the trade.

Agreement was reached by all parties to buy out the original owners Hayward's, and form the Meat Trade Company. Among members proposed for the board were two prominent trade personalities, William Coggan and close associate J.B. Buer who had, among other business interests, a butchers shop in Malden Road, Kentish Town. William played an important part in the success of the journal, not least arranging for new printers to be engaged, Trustlove & Bray of Norwood, an association which lasted seventy years. He was eventually appointed Chairman of the board, serving from 1922 until 1928. The Meat Trade Journal, was originally printed weekly in old Lilliput Victorian style priced 2d, but is now A4 size and currently published fortnightly

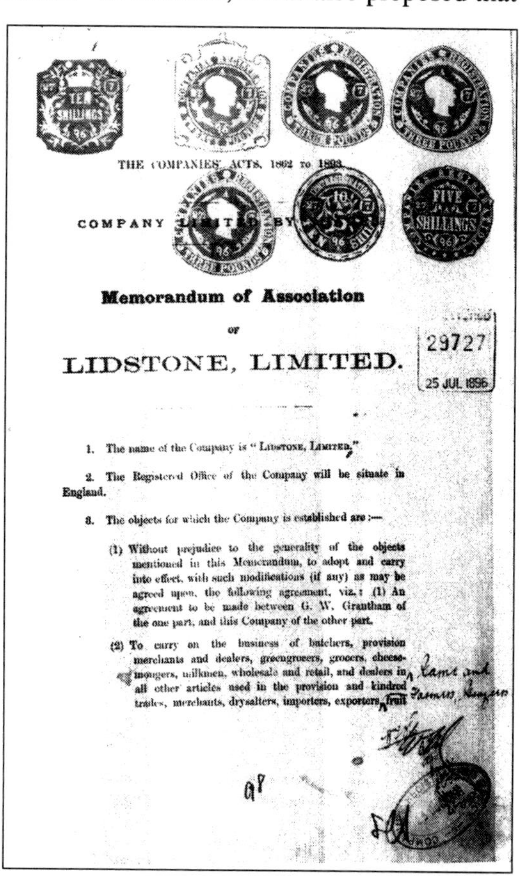

TEN SHILLINGS

FIVE SHILLINGS

THE COMPANIES' ACTS, 1862 to 1893.

COMPANY LIMITED BY

Memorandum of Association

OF

LIDSTONE, LIMITED.

29727

25 JUL 1896

1. The name of the Company is "LIDSTONE, LIMITED."

2. The Registered Office of the Company will be situate in England.

3. The objects for which the Company is established are :—

(1) Without prejudice to the generality of the objects mentioned in this Memorandum, to adopt and carry into effect, with such modifications (if any) as may be agreed upon, the following agreement, viz.: (1) An agreement to be made between G. W. Grantham of the one part, and this Company of the other part.

(2) To carry on the business of butchers, provision merchants and dealers, greengrocers, grocers, cheesemongers, milkmen, wholesale and retail, and dealers in all other articles used in the provision and kindred trades, merchants, drysalters, importers, exporters, fruit

Figure 103: Lidstone Ltd, Memorandum of Association, 1896.

and available on subscription. The irony of its in-house name, the 'Butchers Bible', a title that would not have escaped the attention of Williams's grandson Donald and future Archbishop of Canterbury yet to come.

For some time William had been formulating plans to establish a multiple retail meat company as a going concern from among his close associates currently trading as independent butchers and on the 25th July 1896 Lidstone butchers were incorporated as a limited company, by the merging of twelve meat and provisions businesses to form the basis of the company. Principle participants under an agreement contained in the memorandum of association were Chairman William Coggan and Managing Director George William Grantham, a Camden Town butcher; other participants Messrs Charles O'Hara, William O'Hara, James Collingwood, and Edward Wright that with the exception of Coggan were all based in St Pancras. The inventory comprised of single and double unit retail butchers shops, two coachouses and stables, together with various assets. The main premises with branch shops were situated in the areas of Kentish Town, Camden Town, Hampstead, Kilburn, Belgravia and Warwick Square. A clause in the agreement prohibited each individual from similar business undertakings for a period of ten years within a radius of five miles from each of the premises sold. This excluded a poultry business owned by Charles and an Italian Warehouse and Cash Store owned by William O'Hara. All the named butchery businesses were licensed for slaughtering under local authority regulations.

Of minor detail there were several butchery firms using the Lidstone name and one Messrs Lidstone & Co, 110 Bond Street, purveyor to Queen Victoria felt it necessary at this time to air their obvious annoyance openly with a newspaper advertisement to inform the public they should not be confused with others using their name, and had no other place of business than the stated company address.

The Lidstone company created by W. Coggan had a trading policy throughout which sold a diversity of meats and poultry applicable to the social class of a particular area. The idea of a corporate identity with instantly recognisable logos, livery and regimented window displays had not yet become widespread within the retail sector of the meat industry; although retaining existing former shop titles rather than risk loosing customer loyalty was practiced. Hence Lidstones traded under the Markey banner at 93 Haverstock Hill and 134 Drummond Street, in the nineteen fifties. Prior to this Arthur Augustus Markey became involved in two brief partnerships with butchers S.H. Hurst and H. E. Anstee both dissolved in the nineteen twenties, and Harold Anstee later traded under his own name in Belsize Road.

As so often happened, Camden Town stole the commercial limelight, and Lidstone Butchers are best remembered in the Camden borough for their large double fronted shop at 19-21 Parkway-formerly 75-76 Park Street, and for a period the registered company offices until relegated to North West district office following company expansion.

A matter of importance to the trade during 1901 was the early closing bill, and a select committee of the House of Lords had been appointed to hear evidence from interested parties in relation to proposals to regulate shop hours. As Chairman of both the leading trade organisation in London and his own company, the evidence of William Coggan carried considerable weight, and on Monday 29th April he addressed the committee stating, 'It is generally agreed that the number of hours many shops are kept open far exceeds the requirements of the general public', (this referred to week days), 'and to prevent unscrupulous traders taking advantage, Sunday Trading should be prohibited.' The private slaughterhouses were still sacrosanct and explicitly exempt from the Butchers London Trade

Society proposals. The Early Closing Bill did not become law until 1912, and as it transpired was a mish-mash of government and local authority exemptions and loopholes which satisfied neither the trade nor the public.

Figure 104: The Coggan Family 1901.

As the nineteenth century passed into history, and approaching his fiftieth birthday, William and his wife Elizabeth were living a happy and contented life. His achievements thus far had been considerable, Chairman of a thriving meat company, together with financial interests in hide and skin production, slaughtering, agriculture including stock farming, cereal and animal foodstuffs; among them The Anglo- Rumanian Produce Company, Limited.

In meat trade organisations he was equally energetic, President of the National Federation of Meat Traders, Vice President Elect of the London Butchers Trade Society and Chairman of the Board of the Meat Trade Journal. A man of boundless energy, in 1898 William was elected to the council of officers for the Smithfield Club, a post he held until 1910 and during his tenure officiated as judge at many of the Smithfield Shows.

He was rightly regarded as the trade's most powerful advocate, though his charitable work for the welfare of trade employees was less publicised. Such was the genuine high regard for William Coggan within the trade and beyond, on Monday 13th May 1901 at the Hotel Cecil, 76-78 The Strand, a public banquet was held in the grand hall to honour this respected trade figure. The event was organised by the London Butchers Trade Society, as simply and movingly stated by the Society because "He had done so much for the meat traders of London'. In truth, the entire meat industry and grateful members of the general public wanted to show their appreciation, a fact acknowledged by the large attendance of representatives from all social levels and regions of the country.

The guest list, a veritable who's-who of the meat industry, government officials, civic dignitaries in abundance including three mayors from the London boroughs of Fulham, Lambeth and Southwark, others of particular interest extracted from an extensive list - all retail butchers were J.B. Buer, J. Rayner, Kentish Town; G.W. Grantham, Camden Town, and Thomas Gurney Randall and William Cole from Hampstead. The presentation and testimonial speech was given by Sir Albert Rollit, Member of Parliament for the borough of Islington.

His underlying theme was predictable, less restriction and restraints, the universal war-cry of business men through the ages interwoven with a roll call of perennial problems still troubling the industry; just compensation for loss of meat or cattle through disease or death, reduction of exorbitant freight charges by some railway companies and a uniform rate countrywide. The prohibiting of Sunday trading, and a reduction in the number of weekday shop hours etc. etc.,

Some government proposals were defeated or their impact reduced, as in the Meat Marketing Bill, but slaughterhouse rights had been secured, a direct result of the untiring efforts and leadership of Mr William Coggan, of whom it was said 'Only someone of your consummate tact and inflexible integrity could have successfully coped with'. The presentation gifts consisted of a magnificent solid silver tea and coffee service, a pair of candelabra with Corinthian columns, an inscribed silver salver which read in part 'From representatives of all sections of the London meat trade.' To complete the gift, leather bound volume containing the presentation address, and over four hundred and fifty signatures of subscribers. All the silver items were produced by Mappin Brothers, later Mappin & Webb, the name synonymous with fine craftsmanship.

Acknowledging the testimonial, William demonstrated why he so richly deserved this recognition; beginning by refusing to take personal credit for his achievements and placing the responsibility for any success squarely on the shoulders of family and colleagues. In relating a story from childhood we have some understanding of his guiding philosophy. The first two shilling piece he had from his father was given to him for repeating the 22nd chapter of the Book of Proverbs, which began "A good name is rather to be chosen than great riches, and loving favour rather than silver or gold"; that was the secret, he said.

Souvenir

of

COMPLIMENTARY DINNER

and

PRESENTATION

to

Mr. Wm. Coggan,

Chairman London Butchers' Trade Society, President National Federation of Meat Traders (Incorporated), &c., &c.

HOTEL CECIL, LONDON, 13th MAY, 1901.

PRINTED AT THE OFFICE OF

"THE MEAT TRADES' JOURNAL," 63, LONG LANE, E.C.

Figure 105: Souvenir Dinner Programme, 1901.

His high moral business ethics proved an invaluable asset when he was instrumental in breaking the price ring operated by the leather and hide merchants against the interest of the meat trade; the prices then offered to cattle farmers were as low as sixpence per hide. He founded his own company, the London Butchers Hide & Skin Co. of which he was chairman in 1912; with fleets of carts visiting farms to collect hides, which were transported to warehouses in Bermondsey, a major collection and processing centre. The leather was auctioned each Tuesday, and on the following Friday the farmers received payments being considerably more than under the old system.

His expertise was again in demand during the Great War 1914-1918, when he was appointed technical adviser for hides & skins at the war office; an immensely important post, as colossal amounts of leather and sheep skins were required to satisfy the armed forces. Always in short supply, members of the Coggan business travelled to South Africa to procure cattle hides, reputedly in return for cooking utensils. The services of Lidstone Horse Slaughtering Co Ltd located at the Metropolitan Cattle Market, Islington were similarly in demand for reasons we need not dwell upon.

It was in 1914 the government introduced the working classes cost of living index containing eighty every day items, only twenty of them non-food. In 1918 Bonar Law, the Chancellor appointed a committee including William Coggan among them, to monitor the index and more familiar today as the Retail Price Index; one of the findings being that estimated average weekly expenditure on meat, other than bacon, in 1918 was six shillings and ten pence. He was an early recipient of The Most Excellent Order of the British Empire founded in 1917, chiefly to recognise service by civilians during the First World War; he was awarded the O.B.E. in January 1918 for services to the leather and tanning industry. A happy and well disserved occasion that was no doubt marred by an extraordinary general meeting of Lidstone Ltd board members convened in the same month at Arthur's Stores 114-120 Westbourne Grove, London.

Figure 106: Arthur Coggan, N.F.M.T, 1937.

At this meeting it was resolved to remove Camden Town butcher George Grantham from his office as director of the company to be replaced by George Coulthurst, butcher and proprietor of the venue wherein the meeting was held. Other changes included the appointment of Arthur Coggan secretary upon the resignation of Kentish Town butcher William O'Hara. There may have been discord, however reasons for there departure are unknown. As residents in Woodquest Avenue, Herne Hill from the late 1890s, William and wife Elizabeth enjoyed a comfortable and comparatively affluent life style; the house name of 'Claudia' from ancient times at odds with their modern and rather expensive black saloon Ford motor car. In 1926 aged seventy-four this respected elder statesman of the meat industry died, the cortege following the traditional horse drawn funeral carriage to Brompton Cemetery numbered in hundreds.

A FATHERS SON

His son, Cornish Arthur Coggan, could not have been unaware his father was a hard act to follow. Preferring the name Arthur he was born on the 17th August 1876, inheriting all the natural virtues of his father, and an even greater interest in public service and community affairs. He was educated at Dulwich College, Clacton-on-Sea, and Montrose College, and at fifteen years old he joined the family business starting at number 4 Motcomb Street, Belgrave Square, the flagship of Lidstone Butchers Ltd. Although educated in the purely commercial and financial aspects of business, and later a Fellow of the Chartered Institute of Secretaries, it would be unwise to suggest he was not taught the practical rudiments of shop life during his induction period at this branch. Indeed, fathers invariably insisted that sons and daughters entering the family business commenced at the lowest rung of the ladder, whatever their intended future position. Notwithstanding that by 1901 Arthur's elder brother Frederick was branch manager at one of the Lidstone shops and younger brother Cyril aged sixteen was butchers assistant at another.

The question of his practical knowledge soon became irrelevant, as two years later aged just seventeen he arrived at the Smithfield Market, unquestionably the hub of the world's meat industry at this period. His father had arranged for Arthur to receive a thorough grounding in the wholesale trade, with business friend John Buer of J.K. Buer & Sons, and the firm of Strong and Cooper. His diligence and aptitude for accountancy not surprisingly ensured an early promotion and on his twenty-first birthday he was appointed company secretary in the family business. In 1917 he became a director, a classic example of following in father's footsteps, except for the absence of nepotism, for Arthur had shown from a young age a phenomenal capacity for long periods of concentrated work; thus deservedly any positions held, either in the meat industry or later in local government, was fully justified.

In 1901 at the parish church of Weston-super-Mare, Somerset he married Miss Frances Sara Chubb reputedly the descendant of Charles Chubb, founder of the famous lock manufacturing company. The couple chose 32 Croftdown Road, leading off Highgate Road, as the family home, employing resident nurse Elizabeth Evans and servant Ada Hill from Worcestershire. They had three children, daughters Nora and Beatrice and a son Frederick Donald the youngest, born on the 9th October 1909, whose chosen vocation within the church would eclipse even the distinguished careers of both his father and grandfather.

Figure 107: Councillor C.A. Coggan and Mrs Coggan, 1912.

Family life, it seems, did not inhibit Arthur's single minded purpose of dedication to either his career or local government. His capacity for work increased, as his accomplishments testify, and with the exception of regular Sunday worship at Kentish Town Parish Church time spent with the family was sparse; a side of family life that still caused sadness for his son Donald in his twilight years Although conversely; the family trait of service to others meant that Arthur gave generously of his time and no doubt financial help to good causes through various charitable organisations.

He was a founder member of the Rotary Club of London in 1911, and past master of St. Pancras Borough Council Freemason's Lodge, which he joined through his work in local government and also served as local magistrate. First entering public life in November 1906 he was elected municipal reform member for St. Pancras Borough Council, the scope and volume of committee

Figure 108: C.A.Coggan, Badge of Office, 1912-1913

work he undertook was extraordinary, from housing to disinfecting stations. From the date of his election his high attendance record was never equalled; his dedication rewarded with the accolade of Mayor of St. Pancras for 1912-13, following the unanimous selection by fellow councillors, the enamelled badge with red ribbon and bar C A. Coggan, J. P. commemorating his term in office is in the possession of the author.

The youngest occupant to hold that office at the time, charming portraits of the new Mayor and Mayoress appeared both in the local press and the Meat Trade Journal. Interestingly, printing blocks especially prepared by the St. Pancras Gazette were loaned to the aforementioned publication, an indication of the high esteem in which the Coggan family were held. The original photographs were produced at the Art Room, 271 Kentish Town Road, situated between Crown Place and Holmes Road.

By the end of his tenure as Mayor, military conflict in Europe was on the horizon and in 1914, exempt from military service due to poor eyesight, Arthur joined the Special Constabulary; his area of responsibility included Smithfield Market; he retired from the constabulary in 1935 with the rank of Chief Inspector. The family moved to 57 Asbourne Avenue, near the junction of Finchley and North Circular Roads after the First World War, and on the 3rd July 1919 he joined his father as freeman of the Worshipful Company of Butchers.

From the early 1920's he switched allegiance from his local parish church to St Martin's Church, Allcroft Road, Gospel Oak, an association both as member of the congregation and church warden, encouraging his son Donald to have organ lessons there. Arthur's working week was a constant cycle of company matters, meetings of various trade organisations and a full itinerary of council meetings which apart from the family business were all voluntary and unpaid. The business of Lidstone butchers, from which all things stemmed, continued to prosper and with the establishment of Mutual Trade Insurance in 1925 founded by Arthur, the Coggan family reaffirmed their position, if it were needed, as one of the most influential names in the meat industry.

There was no sign of Arthur slowing down as he continued to accept new challenges, President of the North London Meat Traders Association in 1929 and emulating his father as President of the National Federation of Meat Traders in 1937. While still in office he married his second wife Nora Booth in July 1938 at St Martin's Church, Vicars Road, Kentish Town. Once again on the outbreak of the Second World War Arthur answered the clarion call, serving on various food control committees and attending conferences at the Ministry of Food. In addition to his other interests, in 1941 he became one of the first shareholders of the Meat Trade Journal. Sadly less than one year on, following an operation, this extraordinary dedicated man died. The internment took place at Highgate Cemetery on 24th November 1942.

ALWAYS OPEN

Figure 109: Archbishop Donald Coggan.

His only son Frederick Donald a future prelate, would serve his fellow man in a different way. An unspoken fact known to the family with certainty from his absolute devotion to the Christian faith that began in his teenage years, a calling that has no closing times and Sunday is the busiest day of the week. Ecclesiastical history and the meat trade may appear strange bedfellows to modern day thinking, but that is exactly what they were one impinging upon the other and religion gaining the upper hand. So much so that not only did the church have a controlling influence on the trade, but also shaped the wider society. Attracted to a religious vocation as a young man, and like his father preferring his middle Christian name, Donald he was confirmed at St Peters Church, Dartmouth Park Hill, NW5 where he worshipped. He was later ordained at Fulham Palace on 30th September 1934, and appointed curate of St Mary's Church Upper Street, Islington, spending the rest of his life as a servant of Christ. Future appointments would include Professor of New Testament Wycliffe College, Toronto, Canada, Principal of the London College of Divinity, Bishop of Bradford, and Archbishop of York.

Figure 110: Family Home 32 Croftdown Road, 2005.

A world away from local high street retailing in which Lidstone butchers played a prominent part and which Donald freely acknowledges had contributed to a superior education and a moderately comfortable life style, though was rarely a topic of conversation on his return visits home. At this level of business ownership it is hardly surprising his father divorced the family from the affairs of boardroom meetings; that is not to imply Donald was not fully aware and sympathetic to the labours of many people less fortunate than himself. He spoke of the long hours of work and poor living conditions witnessed in his formative years spent in Kentish Town and later working in Islington.

Whilst his knowledge of detailed company affairs was undoubtedly scant and would remain so, there may as has been claimed a covert attempt at indoctrination, as Donald was educated at his father's request at the Merchant Tailors School, Charterhouse. Square, EC1, in the immediate vicinity of Smithfield Market, near Lidstone Butchers head office. Whether there were any serious hopes he would one day pursue a career in the meat trade was never openly discussed with his father.

Although courted by the meat trade fraternity while Archbishop of York, and no doubt happy to oblige, he visited the York Butchers Guild on ladies feast night and took the oath of membership; he was among fifty subscribers who gave donations to help finance a short history of the guild. During two overseas tours on official church matters in 1967-70, he quite by chance visited many of the districts that were influential in the development of the frozen meat industry world wide; the towns of Nelson, Dunedin and Christchurch in New Zealand and others in Australia, including drinking tea from a billy-can on a sheep station. In January 1975 he was appointed the one hundred and first Archbishop of Canterbury and in the same year made Honorary Freeman of the Worshipful Company of Butchers. The latter privilege also accorded Her Majesty Queen Elizabeth the Queen Mother in March 1997, the first member of the Royal family to accept the Freedom in the history of the Guild. The Royal connection continued in February 2003 when Her Royal Highness, The Princess Royal acceded to the Honorary Freedom of the Worshipful Company.

Upon his retirement, Donald received the Sovereigns honour Baron Coggan of Sissinghurst and Canterbury in the County of Kent; thereafter with his wife Jean he spent a long period of peaceful and happy retirement reflecting and writing books, before moving as he said "To my cosy flat" in St Swithins Street, Winchester.

In the twilight of his life and free from the restraints of high office, church spokesmen, hack writers after a front page story he spoke candidly to the author of the family business. ...... "My father would set off from our home in Croftdown Road, and later Golders Green, for his offices in Smithfield morning by morning and returned at evening without ever discussing business affairs....... I of course knew the Lidstone shop in Parkway and may have visited there, but as far as I recall, we never visited any of the other Lidstone shops though we were well supplied with meat, ... the memory of those entrecote steaks still makes my mouth water ....Kentish Town holds fond memories for me........".

As a young man he remembered the time when Parkway was known as Park Street, renamed in December 1938 when the occupants were notified by the St Pancras Borough surveyor the road would be given its present name and renumbered. Donald died peacefully on Wednesday 17th May 2000, his place in history assured; his wife Jean (nee Braithwaite) died on 24th January 2005. There exists today 'The Meat Trades Christian Fellowship' for those working in and connected with the UK meat trade, which I am sure, would have pleased him. The importance of his father and grandfather's long association with the meat trade has been forgotten with time and is now almost unknown outside the inner family. The Coggan tradition of meat trading however is far from finished, and to this day descendents of distant cousins are still involved in meat processing, wholesale distribution, and farming.

Returning to Lidstone Butchers Ltd, it is unknown when members of the Coggan family finally relinquished financial interest in the company, a business with a precarious financial background that in 1947 required three-quarters of the share capitol to be written off, and signalled dire consequence for the future had it not been for the intervention in 1963 of two Smithfield market men. From that date Gerald Stitcher and John Silver, future joint-chairmen of the company, began buying Lidstone shares and by 1963 owning 70 per cent of the ordinary and preference shares enabling the long suffering ordinary shareholders to be paid a dividend for the first time in sixty-five years.

Figure 111: Lidstone Butchers Ltd, 19-21 Parkway, Camden Town, c1950.

With an estimated thirty branches together with other trading ventures the company seemed outwardly successful, but from the late 1960s the retailing activities were already in decline. Most noticeable for residents in Camden Town the shop and head office in Parkway, this closed in 1968 when the head office relocated to 67 Kentish Town Road, their branch at 227 Camden High Street remained until 1978.

The Parkway shop was eventually demolished, together with adjacent buildings, to be replaced by a prefabricated unit housing Barclays Bank Ltd; a temporary arrangement during the closure for refurbishment of the bank's branch nearby. Thereafter the unit became a support centre for the home-less under the auspices of Camden Council. By 1975 the Lidstone company retail outlets were facing adverse trading conditions and in the prevailing climate found it extremely difficult to obtain a reasonable profit. The trading situation continued to deteriorate, despite a reduction in unprofitable outlets that in 1978 numbered just eleven shops, a further two shops were sold the following year. The worst was yet to come with the collapse of Smithfield meat traders Gilmore and Partners, of whom Gerald Stitcher was also chairman and among the firm's uninsured creditors the Lidstone butchery group. On the 4th November 1980, came the news Lidstone was in the final throes of existence and members and creditors appointed a firm of liquidator's.

Figure 112: Lidstone Butchers Cart, Frognal, c1909.

The financial transactions were lengthy and complex and would serve little purpose to restate here, other than following the collapse Lidstone Ltd were transferred to New Cavendish Estates Ltd, a property and investment company. As part of the proposal it was agreed Lidstone would loose its name and all meat trading activities would cease. Such are the vagaries of business the registered offices of Gilmore & Partners (Smithfield) Ltd, meat importers & wholesalers, were located at 67 Kentish Town Road and so the Lidstone story ended where it began, in a local butchers shop.

MEATHIST MINIATURE

The observance of Lent, the period of fasting and patience from Christmas to Easter required the virtual suspension of slaughtering, dressing and sale of meat. However under a discretionary system woefully abused, 'bodily infirmity', a royal or religious dispensation to eat meat could be granted, from which evolved the issue of a special licence by the Lord Mayor to hand picked butchers who became known as Lenten Butchers.

14

THE GREATEST SHOP

History records that Napoleon described the English in a derogatory fashion as a nation of shopkeepers, it is therefore unsurprising that our tradition of shop keeping had bred in these islands a self confidence in the ability to feed the nation during times of conflict. In consequence there existed a long held complacency within government and the general public. A view compounded in 1905 by the findings of a Royal Commission convened to investigate war time supplies in the event of hostilities, which in part dismissed any reasonable probability of a serious interference with our supplies. A judgement that was to adversely affect government policy on contingency plans in respect of food supplies reaching these shores during the First World War, of parallel importance to defence both decisive to survival and one powerless without the other.

Food rationing had been contemplated early on in the war but in the corridors of power reluctant senior government ministers were understandably nervous of public reaction. Their apprehension heightened because there was little political motivation and no guarantee such a scheme was feasible. Instead, the government stubbornly clung to the belief a free market was the best method for regulating supplies, a decision that had unforeseen repercussions from the beginning of the war with significant increases in the price of staple commodities, such as meat, milk and bacon. .The retail butchers, especially in London and other industrial areas, finding themselves in the embarrassing situation of running out of meat. Many middle and upper class customers, and bearing in mind the dictum 'hunger is good source', were obliged to purchase grades of meat previously considered poor mans offal. Although generally speaking, the vast majority of the working class public considered these items a normal part of their daily diet.

By June 1916 the position had become untenable and a Board of Trade departmental committee was appointed to investigate the principle causes and recommend such steps as practicable to ensure adequate supplies. The committee submitted their conclusions in a twenty page report in September of that year with recommendations; many of which also implemented in the next war. They found the cost of living of the working classes had risen to about 45%, and wages had increased, but were generally below the rise in the cost of food and other necessities. On this basis the committee concluded there was less distress overall in the country than in a normal year of peace, although they recognised certain classes whose earnings had not increased were hard pressed. The hard pressed were paradoxically victims of the increased purchasing power of the high wage earning families, able to buy basic food items which were in short supply. This in turn began a dramatic upsurge in retail prices that placed poorer families in a catch twenty-two situation, slightly better off than before but still unable to afford the higher prices.

From this moment the government effectively became shopkeepers and had begun the process by which departments and outside groups of increasing size were bought together in the creation of the largest food organisation in the country; the object of which to ration, meaning a fixed and limited amount of something, especially food, given or allocated to a person or group from the stocks available, especially during a time of shortage or a war.

The army unwittingly compounding the situation as billeting and ration regulations of home based units required commanders to provision their troops from local shops, which inevitably increased prices and scarcity of food. The daily ration by 1917, for home based troops: ¾ lb fresh meat or 1lb (nominal) preserved; 14oz oz bread or 10½oz biscuits or flour; 2oz bacon; ½oz tea; 2oz sugar; ¼oz salt. In addition 5½ pence per day cash allowance to be expended solely on messing for the varying dietary of units. That the British military forces were better fed than the civilian population is undisputed, and strenuous efforts were made to supply troops wherever they maybe stationed. In particular thousands of tons of frozen meat from New Zealand and Australia that normally arrived in Britain were designated for foreign theatres of war. The realization amongst servicemen that they were better fed than their loved ones rankled, and together with agitators opposed to 'the rich mans war', fostered feelings of discontent.

A feeling of revolution is too strong a word, but there was certainly an unstable and dangerous situation brewing.

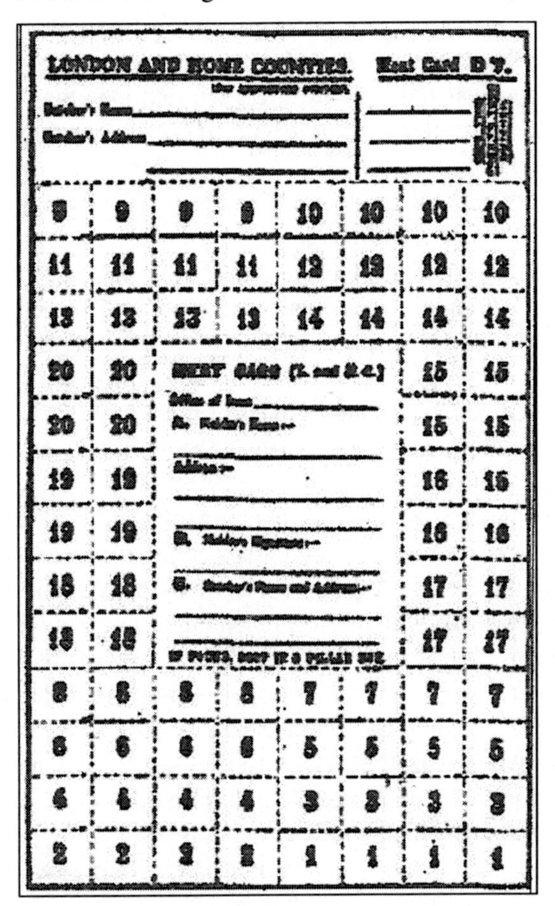

LONDON AND HOME COUNTIES.

9	9	9	9	10	10	10	10
11	11	11	11	12	12	12	12
13	13	13	13	14	14	14	14
20	20					15	15
20	20					15	15
19	19					16	16
19	19					16	16
18	18					17	17
18	18					17	17
8	8	8	8	7	7	7	7
6	6	6	6	5	5	5	5
4	4	4	4	3	3	3	3
2	2	2	2	1	1	1	1

Figure 113: London Meat Ration Card, 1918.

To counter balance the situation the government advised acceleration in the construction of merchant vessels and more dockworkers to reduce congestion at the ports, and to this was added reducing the imports of non essential items. The more radical proposals included voluntary meatless days, restrictions on the retail price of meat and the Government becoming active suppliers of meat by bulk purchase, the latter suggestion implemented by establishing The Imperial Government Supplies Department.

For the patriotic eater and confirmed vegetarian, meatless menu restaurants like Eustace Miles, Charing Cross, Shearn's Fruitarian in Tottenham Court Road, and St George's House in St Martins Lane and The Food Reform, in Grays Inn Road, were already in existence. Where traders made excessive profits it was suggested local authorities should be given powers to open municipal food shops; for women with babies and young children, there was also a recommendation that maternity clinics and nurseries should be used for the provision of milk and meals. The report concluded that seven members of the twelve man committee, under the chairmanship of J M. Robertson, strongly advised the public control of the prices of home produced meat, bacon and milk.

By 1916 dwindling stocks of food had become a highly sensitive political issue, and contrary to the next war they were decidedly muted on stressing the nutritional values of a particular food, especially meat, which might draw further public attention to the erratic supply situation. Divine intervention temporally helped to ease the pressure, with church leaders urging the nation to buy only the minimum essentials, a theme constantly repeated to packed congregations in sermons and proclamations throughout the land.

On the 1st January 1917, a weekly food rationing scheme for each adult and child was implemented beginning with butter and margarine, with sugar, jam, and tea added in quick succession. By years end, on 27th December, control of meat had begun and on 23rd February 1918 a personal rationing scheme was introduced in London and the Home Counties extended on the 7th April to the whole of Britain. All public directives from the Ministry of Food were widely available to trades people through booksellers priced 1d (one penny) or post free from H. M. Stationary Office.

However, for one coupon each adult (children half the amount) could purchase 5d worth of uncooked meat or 10d worth of offal and from 4oz to 16oz of bacon according to cut. The uncooked sausages allowance 6oz or 8oz according to meat content, and Horseflesh or Venison 8oz on the bone or 10oz boneless. Cooked meats allowance ranged from 3oz to 8oz depending on choice, as with cooked sausages, beef, pork, mutton, and edible offal etc. Coupons were not required for black pudding, faggots (savoury duck), and poultry and game were sold by number irrespective of weight for example chickens up to 2 lb one coupon, up to 3 lb two coupons etc. All small birds, wood pigeon, blackbirds and the like were exempt, although all items were subject to adjustments and availability.

There was a sting in the tail for butchers and multiple grocers who sold only pork or frozen meat, as only retailers which sold the full range of fresh meats were allowed to accept customer registrations. As a matter of expediency and profit, many retailers hurriedly added these items to their range of meats or faced the prospect of fewer customers.

The meat rationing scheme bought an immediate back lash from the trade in regard to meat prices, as distribution costs and livestock prices were left out of the equation. The Government were forced to acknowledge that the complexities of distribution, from the farmer's field to the consumers table, could not be solved by merely tinkering with retail prices. It was not until Lord Rhondda, the Minister for Food, decided on complete control, in conjunction with the meat industry, the inequalities within the trade were solved. The St Pancras food traders were concerned that the making of cakes and pastries from wheaten flour was unnecessary and instead the use of flour made from cereal was recommended to the Food Controller.

No amount of government legislation could dictate the outcome of the weather, a bad summer meant poor harvest and that was the end of the matter. The same applied to the rearing of live-stock where nature took its course, intensive farming methods as understood today were still in their infancy; the improved methods of agriculture then for example amounting to simply making greater use of uncultivated land by the rotation of crops, mechanical seed drills and improved fertilizers. All practiced and advocated since the middle of the eighteenth century. A large agricultural Land Army was recruited which included women farm labourers, the beginning of the Women's Land Army.

UNIFORM SCALE OF MAXIMUM RETAIL PRICES FOR MEAT

LONDON & HOME COUNTIES

JOINT OR CUT.	Price per lb. s.	d.
BEEF.		
Topside of Round	1	8
Topside of Round, Best Cut, Boneless	1	10
Silver Side, with bone	1	6
do. do. Boneless	1	9
Thick Flank	1	7
do. do. Best Cut	1	8
do do. Knuckle end	1	6
Aitch Bone	1	0
do. do. Boneless	1	5
Sirloin	1	7
do. Cut	1	8½
do. Rolled, Boneless	1	11
Thin Flank	1	0
do. do. Rolled, Boneless	1	4
Leg and Shin, whole	0	8
Leg and Shin, Boneless	1	4
Suet	1	6
Fore Ribs	1	6
do. do. Boneless	1	9
Wing Ribs, 4 bones	1	8
Long Ribs	1	4
do. do. Rolled. Boneless	1	9
Back Ribs	1	3
do. do. Boneless	1	7
Top Ribs	1	3
do. do. Boneless	1	7
Brisket	1	0
do. Boneless	1	4
Clod & Sticking, with bone	1	0
do. do. Boneless	1	4
Rump	1	8½
Rump Steak, Boneless	2	2
Fillet Steak	2	2
Buttock Steak	2	0
Thick Flank Steak	1	10
Chuck Steak	1	8
Gravy Beef	1	4
Minced Beef	1	6
Sausage, to contain not less than 50 of meat	1	3
Sausage Meat do. do.	1	1
Sausage, to contain not less than 67 of meat	1	6
Sausage Meat do. do.	1	4
Bones	0	2
MUTTON & LAMB		
Leg, whole	1	7
Leg cut, Fillet	[illegible]	7½
do. do. Shank	[illegible]	7½
do do. Middle	[illegible]	10
Loin, whole	[illegible]	5
do. Best end	[illegible]	8
do Chump end	[illegible]	5
MUTTON & LAMB (continued)		
Loin Chops, not to be trimmed	1	10
Saddles	1	5
Shoulders, whole	1	5
do. cut, Knuckle side	1	5
do. Blade Side	1	5
do. Middle	1	6
Neck, whole	1	2
do. Best End	1	6
do. Middle	1	2
do. Scragg	0	11
Best Neck Chops	1	8
Breasts, whole	0	11
do. Cut, Best End	1	0
do. do. Fat End	0	10
do. do. Sliced	1	2
Suet	1	2
PORK.		
Legs, whole	1	6
do. Cut, Knuckle end	1	4
do. do. Middle	1	10
do. do. Fillet	1	8
Hind Loin, whole	1	8
Chump end	1	7
Best end	1	9
Fore Loin, or Griskin or Spare Rib without blade bone	1	8
Hand, with Foot	1	3
Loin, ex. back fat	1	8
Best End	1	9
Neck End	1	7
Shoulder without Hock	1	6
Blade Bone	1	6
Belly	1	7
" best or rib end	1	8
" in slices	1	8
" thin end	1	6
Flare or Leaf	1	4
Back Fat	1	0
Chops or Steaks	1	10
Heads, including Tongue	0	9
Heads, ex. Tongue	0	8
Tongues	1	4
Eye Piece or Face	0	4
Chaps	1	2
Hocks	0	8
Feet	0	4
Tenderloin without Bone	1	11
Pork Bones (excluding Factory Bones)	0	3
Sausage, to contain not less than 50 of Pork	1	6
Sausage Meat do.	1	4

Pickled Pork can be sold at 1d. per lb. above fresh pork prices, but must not be sold as bacon.

MINISTRY OF FOOD.

Figure 114: First World War, Retail Price List.

THE REALITIES

On the home front the war began on a fervour of flag waving, patriotism, and jingoistic rhetoric, encouraging thousand of young men to volunteer for military service. The misplaced idea the war would be 'over by Christmas' quickly dispelled amid increasing numbers of casualties. From day one, contrary to appeals from government and shop keepers, people rushed to the shops in a frenzy of panic buying, long queues formed as harassed staff did their best to cope with demand. The consequence of their actions activating sooner than necessary the fundamental law of business 'supply and demand', and within days wholesalers, especially of imported meat and provisions, had increased prices to retailers who passed them onto the customer. It was around this time that the English gained a reputation for forming orderly 'queues', which incidentally is derived from the French meaning beef tail e.g. queue de boeuf, although ironically ox tail was a scarce item in the shops. The peculiarities of national habits aside, the shopkeepers themselves were not having an easy time.

For the insatiable demand for men of fighting age caused disruption to businesses by the mass exodus of young men single and married eager to join the colours, and in many instances leaving their employer without notice. The large multiples reported mass defections of patriotic young shop assistants; in one instance forcing Eastmans Butchers Ltd, to close 495 shops during a two year period after 1,500, employees', almost all the young men had joined the colours. The sudden absence of so many men from the work place created serious staffing problems until women could be recruited and trained. The majority of women were happy to replace their men-folk and worked in a variety of industries, professions and other trades.

In the independent retail meat shops it was mainly the proprietors' wives and daughters who took over the business while the multiple and others relied on a combination of retired butchers and reduced trading hours. Trading hours were restricted under the Defence of the Realm act 1914, which included a reduction of public house opening hours. The Ministry of Food relied on the local authority to ensure the rules were strictly enforced; businesses large or small were treated equally as provisions merchant John Sainsbury discovered when the company were prosecuted by St Pancras Borough Council. In that being a retailer of rationed food at shops in Queens Crescent and Kentish Town Road, they failed to comply with instructions ordered by the Borough Food Control Committee that such shops should be kept open till 7 p.m. and 1p.m. on early-closing day. The case aroused great public interest because trades unions representing over a million shop workers were currently active in negotiations with the government for a reduction of working hours. In the event, the case against Mr John Sainsbury was withdrawn provided he adhered to current regulations, after the defence had pointed out the order had been made two months after the armistice. The saving of daylight hours by altering the clocks was not instituted until 17th May 1916, the idea reputedly suggested as a peace time farming measure almost ten years before by William Willett a builder living in Chelsea.

From the battle front, after war office press censorship of dispatches was lifted, the national newspapers and illustrated periodicals began to fill their pages with gruesome stories and horrific photographs with headline banners 'The public must know there are no depths of depravity to which the enemy will not descend'. The events portrayed by extras in propaganda films showing German atrocities, with provocative titles like the 'Beastly Hun', many of which recently discovered to have been made on Hampstead Heath. Today viewed as slapstick silent film comedies of a bygone age, yet at the time had a serious intent to stiffen the moral of servicemen and public against the German nation. The encouragement of controlled aggression on the battlefield an obvious necessity in war time, but hostility between the civilian population in the villages and towns of the British Isles was quite a different matter.

The government soon realized propaganda had been excessive and acted swiftly to put the genie back into the bottle; not however before an initial wave of civil disturbances involving German businesses and individuals threatened to get completely out of hand. Most at risk, and the easiest targets, were retail shops particularly pork butchers and bakers, where the names Zwanziger, Konig Eickhoff, Stein, Schulz, Schneider and Hoffman on the facia above attracted harassment and abuse. The Fleischermeister's, who only a short time before were noted for their spiced beef, brawns and sausages now attracted un-welcomed attention of a wholly insidious nature.

The innocuous German sausage, a speciality of pork butchers had to be renamed the Windsor Roll; in some districts called the patriotic sounding Empire sausage.

There was a great deal of animosity in the retail sector and feelings were particularly strong in Smithfield wholesale market with refusal to serve any butcher suspected of being a German immigrant. The disinterest amongst the English race at that period for foreign languages inevitably resulted in confusion and numerous cases of mistaken national identity occurred. Local reaction to war conditions, as reflected nationally, spanned the whole gambit of human nature from base acts of self preservation to self sacrifice on behalf of others. In some instances, the veneer of civilised behaviour completely stripped away, when mob rule took over as happened all over London. In Kentish Town trouble erupted according to reports in the Hampstead and Highgate Express, following the sinking of the "Lusitania" a British luxury liner. The ship was torpedoed without warning off the Irish coast on 7th May 1915 with the loss of 1,198 lives. The more extreme elements of a crowd bolstered by alcohol gave vent to their frustrations, attacked and looted German baker's shops in Malden Road and Queens Crescent during the night of Tuesday 12th May. More shops were damaged and further attacks were carried out on shops in nearby Fleet Road.

To afford themselves some degree of personal safety and peace of mind, people with Germanic-sounding names thought it wiser to completely change or anglicise them. As Louise Eickhoff, daughter of butchers shop owner Frederick Alexander Eickhoff at 2 Flask Walk, Hampstead, reflected later the family surname was changed to Alexander, 'otherwise life would have been intolerable'. It was reported that one shop keeper in Kentish Town, and it maybe said contrary to intention, displayed the following notice written in German in their shop window, 'This is a Ingelische Shopp'.

An intriguing case involved a Mr Ullmann pork butcher in Chalk Farm Road; it seems innocently linked with Ludwig Betz a German born national arrested under the Aliens 'Restriction Order' in Smithfield Market. At his appearance later at the Guildhall court, the 37 year old Ludwig admitted he had not registered and had given the police officer a bogus address. The officer gave further evidence, the defendant had said he was in the employ of Mr Ullmann and was a butcher by trade. His true employment details were never enquired into and remained a matter of conjecture; he was subsequently found guilty and received six months hard labour. The pork butcher referred to in court is recorded in 1915 as John Ulm at 62 Ferdinand Street, his surname probably a derivation of Ullmann. The shop had two previous proprietors with Germanic sounding surnames John Golterboth and Frederick Englhard.

Hostility and discrimination continued for most of the war, erupting into something more sinister whenever enemy atrocities or heavy British losses were incurred. It was reported J Lyons & Co, Ltd, famous for their tea shops and nippy waitresses, had taken advantage to poach customers by circulating scurrilous and defamatory rumours that grocery multiple Lipton Lid had German connections. The firm vehemently denied such accusations in national press advertisements, at the same time announcing their intention to sue for libel.

To counteract the general feeling of disquiet among both English nationals and residents of German origin, lists of name and exemption began to be published in the newspapers.

The grounds for exemption from internment or repatriation of German men and women as stated by the Aliens Advisory Committee: (1) Long residence in the United Kingdom, (2) British born wife or children, (3) Son or Grandson serving or has served in His Majesty's Forces.

The St Pancras Guardians Schools at Leavesden, near Watford proved altogether more constructive from a practical food point of view in January 1916, sending the carcasses of eight pigs and 5 cwt. of parsnips to St Pancras House. This building was previously titled St Pancras Workhouse, and sited in St Pancras Way then known as Kings Road. The complex at Leavesden consisting of accommodation, schools, workshops, gardens and farm buildings was built on a site purchased by the Guardians in 1868 and opened two years later.

The hoarding of food was prohibited under the Food Hoarding Order, and offenders were widely reported in local and national newspapers. Two high profile cases attracted considerable public attention in 1918, through exposure in both the Hampstead Advertiser and Daily Mirror newspapers. The first concerned Mrs Henrietta Lewy of West End Lane, Hampstead summoned at Marylebone for hoarding food. An inspector of the Ministry of Food said quantities of food found at the defendants home were: 46 ¾ lbs tea, 15 lbs sugar, 30 lbs syrup, 42 lbs flour, 18 lbs coffee, and 16 lbs flaked rice. Fines and costs were imposed amounting to £10. In the same year Dr Augustus Henry Cook, of 5 Rosslyn Hill, Hampstead, and his wife, Helen Neville Cook, were summoned at Marylebone for nine contraventions of the Food Hoarding Order, having in their possession or under their control: 87 ¾ lb of sugar, 51 lb of tinned milk, 66 ¾ lb of tinned meats, meat extracts and fish and 138 ¼ lb of cereals and their products. In mitigation Dr Cook said he had practically no knowledge of the store cupboards, and Mrs Cook stated that since the war she had economised and had herself lost a stone in weight. Mrs Cook was fined £20 and the case against her husband was dismissed on payment of £5.11 shillings costs, and the food was confiscated The prosecutions and heavy fines, aside from discouraging the practice, did however perpetuate the jaundiced view commonly held among working classes lower down on the social scale, the better off were somehow escaping the worst of food shortages. These acts of selfishness and the culprit's addresses, both in an affluent area, could only give credence to such a view point. It has been argued that rationing placed the working class on absolute parity with the wealthiest in the land, in this instance this was plainly not the case.

The overcharging by food retailers was an everyday occurrence, a practice difficult to stamp out until the maximum price order came into effect. Thereafter offenders could expect scant sympathy in the courts, receiving punitive fines and in some case a term of imprisonment. One of many habitual offenders was John Best Ltd summoned for selling beef at 5d per lb, in contravention of the maximum price regulations in their shop at 225 High Street, Camden Town. Appearing at Marylebone Police Court on the 2nd August 1919, the defending solicitor said 'the meat was sold without authority by a cutter-up whilst the manager and other shop staff were busy hanging up flags for the peace celebrations. The magistrate Mr Biron commented 'this company has a very bad record and overcharging must be stopped, and he could not help thinking the patriotism of the company would be much better shown by dealing honestly than by hanging up flags'. The company were fined £100 pounds and he imposed a £10 fine on Mr Charles William Terry the cutter-up. Whilst no actual starvation of the population occurred during the First World War, it had been perilously close; however valuable lessons had been learned from mistakes that would not be repeated.

A LESSON LEARNED

In 1938 once again the spectre of conflict with Germany loomed heavy on the horizon and throughout the summer and winter months and into the following year, more in hope than expectation, a succession of government ministers and diplomats instituted political initiatives to prevent such a catastrophe. As a precautionary measure in an atmosphere of near normality under the circumstances, the country prudently and secretly prepared for war.

A preparation of a different kind also happened that year, with the town hall planners and various sub-

Figure 115: R. J. Dickson, 1939.

committees and hoping for a very different outcome. For in that year the Metropolitan Borough of St Pancras, with the able assistance of staff from the borough libraries, mounted an exhibition that can only be described as the wonders of St Pancras, but was officially titled 'St Pancras Through The Ages', a stunning display of artefacts, illustrations and memorabilia with lectures and entertainment. The week long exhibition was opened on Monday 26th September 1938 by the Rt. Hon. Lord Stamp of Shortlands G.C.B., G.B.E. and staged in the assembly room of the new Town Hall, Euston Road. In twelve short months from the opening, the Borough would begin another chapter in its long and varied history in which the whole nation would be engaged in a war that at times threatened its very survival.

For the time being however local newspapers continued the air of normality with news of C & A Daniels store in Kentish Town Road celebrating their seventy-fifth birthday; with the personal attendance of June Duprez and Donald Grey, then currently appearing in the film 'The Four Feathers' showing at the Odeon Cinema Leicester Square. The London Co-operative Society opened a new branch at 153 Fortess Road, Kentish Town, and Mr H. Gordon Selfridge aged seventy-five, the Selfridge department store magnate had retired. While costermongers continued to cause obstructions in the area, ever streetwise when cautioned by police, one trader claimed that he had stopped to pick up a tanner (sixpence), which had fallen from his days takings on the barrow.

On the 3rd September, 1939 local news was relegated, with the announcement by Prime Minister Neville Chamberlain, we were at war with Germany. Fortunately the lesson of the previous conflict had been learned and by 1935 forward planning for civil defence and food control was already at an advanced stage.

Fearful of air raids, places of entertainment were immediately closed, largely affecting the cinema and theatre going public. Disgruntled patrons and a need to keep morale high soon persuaded the government otherwise, and most reopened within weeks, although some did remain closed permanently. The Everyman Picture Playhouse and Odeon cinemas in Hampstead closed temporarily, but the Odeon in Haverstock Hill suffered bomb damage in 1940 and remained closed until 1954. The closure order also covered football matches and other sports events, fates, parades, association meetings indoor or out, in fact any large gathering of people. The Bakers trade exhibition, and other events scheduled to open at the Royal Agricultural Hall, Islington, were cancelled, although Churches and other places of public worship were exempt.

HERE WE ARE AGAIN!

Some of Page's Staff relaxing and building up for the strenuous time Christmas Trade brings with it. It is necessary to be fit to withstand the strain at this Festive Season.

Established in the days when Beer was 2d. per quart

This is now the Largest Meat, Poultry and Provision Store in North London and is famous for PoliteService & Attention

TONS OF TURKEYS

Tons of Dairy Fed Pork. Tons of Scotch Beef.
Tons of Bacon & Hams. Tons of Sausages & Sausage Meat

Page's Stores — one of the Freshest and Cheeriest Places to Visit at Christmas Time!

PAGE'S STORES,
Camden High Street, N.W.1.
Telephone: GULLIVER 4486-7.

Figure 116: Page's Stores, Christmas Advertisement, 1939.

Following several trial runs in different parts of the country to test the efficiency and public response to black-outs, air raid precaution drills including the use of sirens were organised; a nationwide black-out was imposed for real in 1939. The hours of black-out were given in newspapers, on the radio and flashed onto cinema screens and were rigidly enforced in the magistrate's courts, imposing penalties on offenders for displaying lights during the hours of darkness in contravention of the lighting regulations

From accident or wilfully, offenders found to their cost this meant between a five and forty shilling fine and in extreme cases imprisonment. The majority were regularly named and shamed in extensive lists published by the local press. In a one week period in Kentish Town, the list contained over thirty names; a high proportion of these retailers: grocers, butchers, chemists and radio shops, including a drapery store; ironically the best selling item in the store, being blackout material at a 1/- per yard. The Fortess Shoe Co, Ltd, Kentish Town were fined £25 with five guineas costs because there were eight electric lights burning in the shop window on a sunny November day. Of little importance under the prevailing circumstances, barely twelve months before the public lighting committee of St Pancras Council had submitted plans for improved lighting covering seventy-two miles and affecting 584 streets in the Borough. In war time winter, milkmen were forbidden to begin their deliveries before 7 am in the morning, a regulation it would be have been wiser for one dairy owner in Malden Road, Kentish Town not to ignore as he was found guilty and fined £5 with costs. On one occasion confusion reigned when two London local authorities implemented differing black-out times which led to shopkeepers facing each other across the high street with blazing lights on one side and in total darkness on the other; a monumental blunder reported by every national newspaper in the country.

The seasonal trade advertisements in the local press approaching Christmas 1939 seemed eerily unreal, mixed as they were between offers to construct your own family air raid shelter or convert a basement to withstand anything but a direct hit. Many of whom had traded through a similar crisis in 1914 and no doubt had faith in their ability to do so again

As almost nonchalantly, the likes of Kentish and Camden Town butchers shops Kimbers, Dickson, Pages and Fenns grocery store and others advised customers of their large selection of fresh poultry and argentine meat. It so happened shortly after the G. F. Kimbers advertisement appeared in the St Pancras Chronicle, chilled Argentine imports were suspended. The trade ancestry of advertiser Robert James Dickson stretches back to 1885, as D. M. Dickson & Sons then sharing a corner location with the Adelaide Tavern, facing Haverstock Hill. The business under the proprietorship of Mrs Dinah Marie Dickson who laid the

Figure 117: G. F. Kimber & Son, 1939

foundation for a successful future expanding to four shops extending into breeding prize winning cattle and poultry.

War or no war; undaunted the advertising community found fertile ground in the pre-war nursery of material possession Britain. Gifts for Christmas were much in evidence, books, music, toiletries and clothes, and for that special occasion a Eugene permanent wave price 21/- at ladies hairdresser, Maison Hilda 99a Kentish Town Road. For house proud ladies of leisure used to a daily, the Hoover Company gave reasons for buying a vacuum cleaner at pre-war prices, reminding readers that dust, dirt and germs take their toll and there's bound to be a shortage of servants, come what may. If all this became too much, the store with the west end reputation, Bowman's established 1864 in Camden Town, invited prospective customers to spend the long black-out nights in the comfort of their exclusive furniture. Alternatively, readers were encouraged to leave their worries behind them at Victoria Coach station and enjoy a prearranged Christmas programme of travel excursions. In a play on words one American meat company advised 'Armour your larder with Armour's corned beef', while sausages, newspaper readers were told; taste better for a dash of H.P. sauce. For those on the home front the first Christmas at war ended on a feeling of optimism, however irksome all the regulations and preparations maybe, perhaps things would not be too bad after all. The feelings of optimism and well stocked windows were short-lived, as the so called 'phoney war' ended amid a rapidly deteriorating situation as the full horrors and deprivations of total war soon became apparent.

NEW PARTNER

The responsibility of feeding the nation was given to Frederick James Marquis, 1st Baron Woolton, created a peer expressly to fill the post; he chose the title from the village of Woolton near Liverpool, and former managing director of John Lewis department store in Liverpool, unconnected with the London based firm. He was plucked from the backwoods of political obscurity and appointed Minister of Food on 4th April 1940 with a staff of 3,500. On perceiving the magnitude of the awesome task confronting him, he himself said 'I suppose I am going to run the greatest shop the world has ever known', to which one might also add and the greatest advertising agency. In the ensuing years, as Lord Woolton, Minister of Food until November 1943, he became a household name and his shop staff had grown to 39,000, as he referred to his staff at the ministry. His guiding principle was to ration nothing, however scarce, until there was enough to go round and then to ensure that this ration, however small, was always honoured. From the onset, Government policy was heavily biased towards an agricultural policy with the production of crops, potatoes, cereals and dairy produce, milk, butter and cheese as opposed to rearing livestock. To this end a cultivation of land order was issued in 1939 and agricultural executive committees were appointed for each county.

Our main adversary again during the war, Germany, took the opposite view, concentrating their efforts on increased meat production, which yielded more protein. The British government's position was not intended to infer we were about to become a meat-less nation, on the contrary extensive pre-planning had taken place to see this did not happen; a mammoth task in which the entire meat industry of this country would be involved. The food defence plan department, as it was named, was activated on the outbreak of war and became the Ministry of Food, each branch of the food trades responsible for it own commodity.

For the meat trade initial effects were almost immediate, as all wholesale markets were closed to buyers with the exception of un-rationed meat items. The largest, Smithfield Market was dispersed to pre-arranged depots at Islington, Earl's Court, Lewisham and other locations, all under the direct control of the London Meat Supply Association. At the same time eight regional Wholesale Meat Supply Associations became operational. The allocation of meat from these associations was regulated by retail butcher's buying committees, who then allocated supplies in their areas of control. All meat, wherever it may be stored, was requisitioned by the Meat Importers National Defence Association and dealers required to obtain a licence. All slaughterhouses were subject to compulsory takeover, thereby safeguarding home produced meat supplies. A newly created company, The Bacon Importers National Defence Association Ltd. regulated the supply and distribution of bacon.

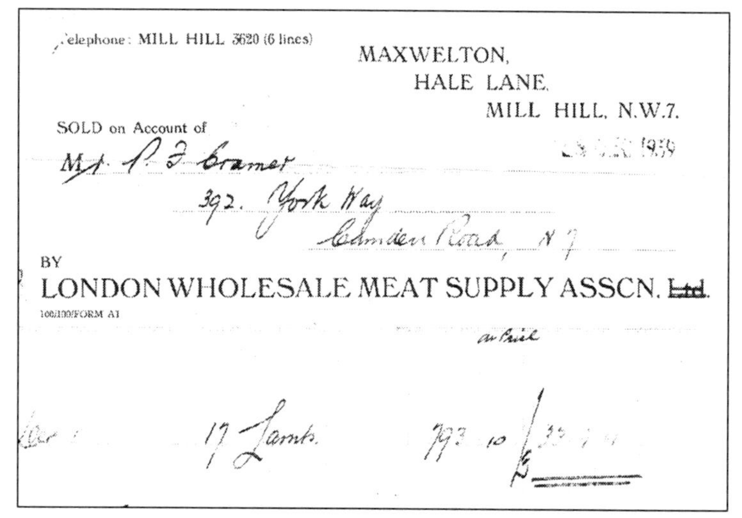
Telephone: MILL HILL 3620 (6 lines)

MAXWELTON,
HALE LANE,
MILL HILL, N.W.7.

SOLD on Account of
Mr P. J. Cramer
392. York Way
Camden Road, N 7

BY
LONDON WHOLESALE MEAT SUPPLY ASSCN. Ltd.

17 Lamb. 79 10

Figure 118: London Wholesale Meat Supply Association, December 1939.

The manufacture and sale of canned meat to wholesalers was prohibited except under license, albeit there were certain specified exceptions. In October 1941 the Ministry of Food issued a press statement that control of meat supplies and distribution was complete, the meat industry was now subordinate and subject to directives from the Government. As for canned meat, butchers who had dealt in these products before the war were allocated rations, similarly all shops and cooked meat dealers who met this criteria. It was considered unworkable to treat pork butchers as a specialised trade, their only options being to change to general butchers, become bacon curers or close for the duration. The final link in the domestic supply chain before consumers; the retail butchers were allocated quantities of meat equal to their sales just prior to the beginning of the conflict, later adjusted to numbers of registered customers and casuals, merchant sailors and forces personnel on leave or billeted.

The Governments aggressive food supply advertising to the general public and 'our butchers' in the autumn of 1939, exuded confidence others were not totally convinced. In particular bacon, an essential ingredient of the 'English Breakfast' that with 'Tea' considered vital to morale and hence the war effort. Once again in times of need old ideas were dusted off and pressed into service, in this instance the reintroduction of 'Macon' a cured or smoked mutton substitute for bacon. The article deemed suitable for breakfast, brunch and Kickshaws the latter meaning a fancy dish, tidbit or delicacy. With bacon already in short supply butcher Philip Cramer, York Way decided to promote the Macon product in his shop, and for this reason purchased and had processed mutton carcasses. As he relates; 'It sold but was not popular', a view expressed by butchers and eating establishments nationwide, and acknowledged in 1940 by a government spokesman with the announcement the Ministry of Food production of Macon would cease.

With regard to imports, to economise on space in the holds of refrigerated vessels, sheep carcasses were telescoped one half inside the other and beef arrived de-boned. So successful was this space saving measure that the maximum cargo tonnage some ships could carry was reached before the holds were full. Hardly credible at such time, boneless hindquarters of beef were pronounced not suitable by London butchers, because housewives had difficulty in identifying the meat. The American boneless beef in particular arrived in such a condition it was difficult for even experienced butchers to tell one cut from another. One of many, butcher George Price with premises in Seaton Place, Camden Town, pleaded guilty to exposing for sale imported boneless beef above the permitted maximum price. In mitigation, the defendant had mistakenly confused one cut of boneless beef with another of a higher price. The magistrate, declaring his sympathy with the problem, dismissed the case on payment of two guineas costs.

FAIR SHARES

The cornerstone of the government domestic policy for fair and equal shares of available goods was The Goods and Services (Price Control) Act 1941, empowering the Board of Trade to make orders fixing maximum charges for services and goods. This Act amended and supplemented the previous Price of Goods Act introduced in 1939, which made it unlawful "to sell, agree to sell or offer to sell any price regulated goods which exceeded the permitted price". This replaced an earlier Act, under which a food order was issued, prohibiting meat being sold above prices charged at the end of August of that year, to prevent profiteering on a wide scale that undoubtedly would have occurred. A temporary measure until the Maximum Retail Price for Meat 1941 order came into force, when shop posters were printed detailing the various joints and cuts of meat and maximum prices; the list came with a legal requirement to be displayed in a conspicuous position.

Figure 119: Ministry of Food, 8th March 1940.

It was now the turn of the general public and National Registration Day was announced for 29th September 1939, based on the census. A national register recording particulars of every citizen was compiled and the information used to issue Identity Cards and Ration Books. The people of the Borough of St. Pancras, motto "With wisdom and courage", of which Kentish

Town formed part, faced the trials and tribulations of war, as did the nation with a wry smile and dogged determination.

In the food office on the ground floor of what is now the Borough of Camden Town Hall; the clerical staff recorded the constant changing personal circumstances of ration card holders, thousands of whom were continually on the move. The ration book counterfoils recording their chosen food retailer were registered or cancelled, and dispatched to the nearest local food office. Several types of books were issued, for adults, children under six years old, travellers and for seamen. The books renewable each year and issued from designated centres on specified collection days and strictly adhered to based on the alphabet A, B, C and D were called one week and so on.

The disruption caused by the bombing meant some odd venues at times, the Royal Veterinary College in Royal College Street, which also doubled as a re-housing centre for persons displaced as a result of the air raids. A staggering sixty million changes of address were recorded within the six years of war, amongst the thirty eight million civilian population. Just how essential their work was and the absolute importance of the ration book is starkly obvious from an incident described by Marlene Crilly in 'Blitz Kids', a heart warming account of how one Kentish Town family faired during the blitz and early post-war period. 'We were dug out from under glass and debris with no more than cuts and bruises to complain of but Mum insisted on trying to find our ration books among the debris and it took the neighbours some time and eventually, physical force to stop her'. The Board of Trade created a service to provide 'utility furniture' for families who suffered during air raids, and young people wishing to set up home. To publicise the service, St Pancras Chamber of Commerce organised a three week exhibition of the furniture in the showrooms of the Gas Light & Coke Company in Camden High Street.

On Monday 8th January 1940 food rationing became a reality, and as transpired an inseparable part of daily life for the next fourteen years, with its own vocabulary as monotonous as wartime queues. Shop notices stating 'Sold Out', 'Only One Each', 'Awaiting Delivery', 'Coupons Required', interspersed with well intentioned, but insufferable homilies, 'Keep Your Chin Up', 'It Will All Be Over Soon' and 'Waste Not Want Not.'

On the first day bacon, ham, sugar and butter and a few weeks later in March, meat was included on the list to the value of 1s. 10d. per week for each adult and child over six years old, the allowance reduced to 11p for children under that age. For fairness and practical reasons the allocation was based on value and not weight; the adult ration cut still further as the supply situation worsened. It was left to each adult to decide how the allowance was managed and the butchers to fairly interpret the rules. It has been calculated, at the prevailing war-time prices, the average allowance was seven ounces of meat and two ounces of corned beef per week.

The sale of corned beef was restricted to butchers shops, as Thomas Buttling working in his fathers shop in Queens Crescent, Kentish Town has reason to remember: 'I vividly remember, when still at school but helping at the shop on Saturdays, spending all morning cutting up 6 lb tins of corn beef for the customers few pennies worth each as part of the meat ration'.

In the second year of hostilities America came to our aid with the passing of the Lend Lease Bill, and the arrival of the first food ship, bringing an immensely palatable if not regular supply of food stuffs.

For off-ration items, food retailers often operated an alphabetical system based on the customer's surname, working their way through the names as and when supplies permitted. The items such

Figure 120: Food Facts Poster No 76, 2nd January 1942.

as sausages, offal, poultry, fish, rabbits, game, goat, and horse meat were un-rationed, but still subject to supply and price control. In addition the points scheme in operation from early on whereby every ration book holder received sixteen points, later raised to twenty, to purchase in any shop the items of their choice The points system allowed flexibility to counteract disruption in home and imported supplies, by increasing or decreasing the number of points required for each product, depending on available stocks; included in this category a variety of imported tinned meats in the form of luncheon meat, sausage meat, tongues, meat loaf, stewed steak, meat roll and rabbits. For commercial use, dried beef and mutton in powder or compressed blocks was imported for makers of pies, cooked meat dishes and restaurants. The words 'if available' being of monumental significance throughout the war. All reinforced by a vigorous Ministry of Food poster campaign including a series 'The Butcher Says' advising the housewife on meat recipes and other related subjects.

The retail sale of horse meat for human consumption, Horseflesh (Control and Maximum Prices) Order, although permitted was subject to specific emergency regulations. "No person shall sell or offer to sell or expose for sale or deposit for the purpose of sale or buy or offer to buy any horseflesh for human consumption, in, upon or about any premises where meat other than horse-flesh is sold or offered or exposed for sale for human consumption, or where horseflesh not for human consumption is sold or offered or exposed for sales". Wholesale dealings were subject to licence, and the manufacture and sale of meat products containing horse flesh were prohibited. There are countless stories of deception regarding this species of meat that have passed into folk law. The number of incidences grossly exaggerated, but newspaper reports of court proceedings verifying abuses and adulteration of meat products were not uncommon.

Specifically targeted with some justification were sausages, at one period rumoured to contain horsemeat and sawdust. The butt of many war-time comedians, with jokes about Hitler's secret weapon, bags of mystery, shamefully adulterated at times that included offal, a full exposure of the ingredients like Pandora's Box is better left unopened. The war time content of pork or beef sausage and sausage meat as prescribed by the Ministry of Food in July 1943: 37% of meat, 7% of Soya and 55% of filler or national flour added water and seasoning or flavourings.

Worst still, for a nation of meat eaters, the promotion of the vegetable sausage together with another item, tinned soya meat products with a limited but obvious appeal to the vegetarian. Vegetarianism is of course complete anathema to meat traders, and early on in the war the London

Vegetarian Society had tried in vain to have extra butter rations in exchange for their bacon and ham coupons. Another extremely unpopular item was whale meat, tinned or frozen, in all its guises virtually indigestible and strong smelling and definitely not recommended for the feint hearted. The un-rationed fish was hard to come by, as the late Wendy Trewin remembered during her lunch break, rushing to Mac-Fisheries branch in Hampstead Village in search of fish rumoured to have arrived, only to be disappointed at the slopping counter filled with tins and packets of bread crumbs. Surprisingly bread remained un-rationed throughout until the post war period.

The well known ABC bakers and others accustomed to lengthy early morning queues for hot bread took the decision in their stride. The Aerated Bread Co Ltd, given its proper title began their commercial life in 1862, the year of incorporation. The head office was in Eastcheap House, 24 Eastcheap, London, and the company title aerated bread encompassed the method of production by a Dr Dauglish, wherein all the nutritious and digestive elements of the wheat are preserved in the loaf. It was also claimed to be the only system by which the mixing, kneading and moulding were done in hermetically closed vessels. A confectioners and light refreshment a contractor, the first bakery was sited in Islington until relocating to Soho in the early nineteen hundreds. By 1923 amassing 150 branch shops, one at 202 Kentish Town Road, and 250 tea rooms in addition to a factory built in 1930 and sited in Camden Road. The firm being absorbed into Associated British Foods Ltd by consequence of a take-over in 1955 by Allied Bakeries Ltd. It is believed the parent company ceased trading in the early 1980's, and subsequently the ABC bread company.

The waste and misuse of food was considered a cardinal sin and also illegal, even feeding street sparrows and pigeons; imports of animal food stuffs had been drastically cut because military requirements took priority on all shipping. A vigorous poster campaign 'If they starve we starve, save your scraps of food' drove home the message. Local councils provided a collection service for kitchen waste by sighting communal receptacles (pig food and bone bins) in streets and roads for recycling into animal feed. It was reported that zoo animals that had died from natural causes were also utilized.

The launch of the 'Dig for Victory' campaign, with the help of voluntary bodies like the National Allotments Society, the Women's Institute and Land Settlement Association, proved a tremendous success between 1940-45, with an estimated eight million tons of food produced from domestic sources. The organising of pig, rabbit, goat, poultry, bees and vegetable clubs were actively encouraged by the Ministry of Agriculture, and local housing authorities and private landlords

Figure 121: Home Curing of Bacon and Hams, Edition 1945.

officially asked to relax restrictions on poultry, rabbit goat and pig keeping. The cottagers pig was once again seen in towns and cities, many in makeshift sties amongst the bombed ruins and allotments. At the time local pig club news appeared in a November 1942 issue of the St Pancras Chronicle, over 4,000 clubs countrywide had been registered. Several managed by St Pancras fire brigade units and other auxiliary service personnel, one of many in Oak Village, Kentish Town operated by patriotic members of the public. The town and country domestic poultry keepers accounted for quarter of the total egg production, and the rapidity of the wild and domesticated rabbit population to reproduce off-spring, an unwelcome trait in peace time, during war time became a positive asset. Special evening trains were diverted via the West Country to collect thousands of rabbits for delivery to London each day. These supplemented tinned rabbit sold under the points system, which had been stock-piled from 260,000 tons of frozen and canned meat imported in 1939-40 from Australia.

Not always seen in the best culinary light by housewives, rabbit with onions were a means to an end for one disreputable shopkeeper. As in the Great War, food became a second currency and retailers took upon themselves discretionary powers far beyond those legally imposed by the regulators, for example potatoes only sold with greens or onions. Of the three commodities, onions were the scarcest and the most sort after and therefore possessed an exorbitant value in terms of bartering and sale; which is aptly demonstrated by a transaction between an undercover food inspector and a shop assistant at a provisions shop in Junction Road. The inspector, at the instigation of a customer complaint, tried to purchase an onion but was told by the assistant these were only sold on condition customers bought a rabbit as well. In addition to the assistant imposing an illegal condition of sale, the offence was compounded by selling both the onion and rabbit at a price exceeding the maximum price allowed. For this blatant piece of food black mail and extortionate pricing the owner was fined 40 shillings and three guineas costs. One newly married housewife had plenty of onions, Mrs Gladys Saunders recalls in the summer of 1942 waiting excitedly for her first anniversary present, promised by her husband away at sea. 'At last the postman arrived at our house in St Albans Road, and he was carrying a large flat cardboard box with Egypt stamped on it. I expected to find some exotic dress, only to find he had sent me onions. I was fuming but it was my own fault as I had mentioned to him in a letter we had been without onions for months. I do remember going to J. Stebbings greengrocers in Raydon Street without any luck. Anyway I made a bacon and onion roll and gave some of the onions away to relatives and neighbours. Mind you, we had many a laugh over it when he came back home'.

A deluge of Government food propaganda via information films and food flashes, wireless broadcasts, books, posters and leaflets did its best to convince the population that necessity really was the mother of invention. If the housewife only persevered, that tired old breast of lamb could be transformed into beef stroganoff. Even packed lunches were not immune, and food facts information sheets covered these too. The local papers serving the borough provided a valuable service with public announcements and good advice on every conceivable subject. The St Pancras Chronicle weekly peace time column 'World of Women' continued throughout the duration, but pre-war items on how to modernise your kitchen and spring hats were not surprisingly discontinued; instead regularly reminding readers in a reference to preserved fruits, 'When there's blossom on the trees there's naught on the table', followed by recipes for Plum Pudding, Quick Dumpling, Rhubarb Poly and Baroness Pudding which hopefully were an improvement on Woolton pie. How many housewives had the ingredients, opportunity or inclination to try these and a thousand and one other recipes we shall never discover.

The sweet dishes were of lesser concern however, as home life since the turn of the century demanded the main meal of the day with the possible exception of Friday, should include meat of some description on this the public was adamant. So entrenched was the Sunday roast and other home prepared meals containing meat, the Ministry of Food found it necessary to issue a Meals without Meat information leaflet number 29, bluntly describing 'seven appetising meatless meals' as an alternative. These, contrived mock or imitation food dishes to disguise the absence or substitution of the main ingredient invariably meat, became an art form employing the ingenuity and imagination of a whole government department and a legion of cookery writers. Diners everywhere, when confronted with one of theses inspirational dishes mock duck, goose, rissoles, venison and others, knew they were about to make an enormous leap of faith into the unknown.

To bolster moral, Mrs Winston Churchill visited St Pancras in October 1941 and was received at the Town Hall by the Mayor Evan Evans, who reported afterwards the wife of the Prime Minister was exceptional, pleased with everything she had seen. The role of women in the Second World War was far more varied than in the previous conflict, although primarily they were still called upon to work on the land and in factories to replace thousands of men conscripted for the armed forces. Their employment in the retail meat trade, aside from driving, was patchy and much depended on the individual circumstances of each shop. The practice in multiple butchers and grocers who sold meat was more generalized, although there was a definite attempt to preserve the female dignity by insisting, as Sainsburys had done, they must not be referred to as butchers. A laudable sentiment, that went hand in hand with femininity and the irritating 'Make do and Mend' culture as a consequence of clothing coupons. The situation compounded by the protective clothing, consisting of turban and coverall, worn by thousands of women on war work. A matter of minor importance, but irksome nonetheless, the refusal by Sir Andrew Duncan president of the Board of Trade to allow handkerchiefs to be made coupon free; the official allowance, four small hankies for one coupon; clothes rationing eventually ended in 1949 To keep the workers happy and busy in factories, the lively war time tune 'Calling all Workers' composed by Eric Coates who lived in West End Lane, Hampstead, introduced the twice daily radio programme 'Music While You Work'. For their nearest and dearest holding the fort at home, the familiar signature tune 'In Party Mood' meant 'Housewives Choice', a much loved programme presented by George Elrick. There was a touch of unintentional black humour given the title tune under the prevailing war time circumstances.

The public were continually updated, via official information, concerning all the various rationing schemes with posters such as 'Meet Your Ration Book'; although everything depended on the supply situation, and so rationing was inevitably subject to constant adjustments and alterations. All food retailers were under enormous pressure, particularly the butchers. It is impossible to overstate the importance of meat, both dietary and physiologically at this period, for that reason customers who also had

Figure 122: Woman Butcher, Sainsburys, 1940s.

their own private worries were extremely sensitive to the slightest indication they were being treated unfairly. The butcher had to careful in everything he did to avoid an ugly scene or be faced with a hostile crowd of women customers. One flash point arose from a butcher who accidentally weighed a customers corn beef with two sheets of grease paper instead of one, and tempers flared when another attempted to weigh a cut of pork with the trotter attached, a pre-war habit that was now illegal as regulations stated trotters could not be sold attached to legs or hands of pork.

It paid to be nice to your butcher but this did not extend to the activities of one local Kentish Town butcher, when a ten shilling note inducement for extra meat rations was seen to fall from a ration book. The embarrassed butcher caught in the act and challenged by other customers in the shop, miraculously produced a little something extra for each customer. This was presumably to buy their silence, as in an order made under the regulations, shopkeepers were forbidden to supply any article of food in excess of customers' normal requirements

On the other hand there prevailed amongst magistrates some understanding of the difficulties that butchers and other retailers traded under an avalanche of regulations at the beginning of the war, and subsequently absorbing the various amendments and supplements that followed throughout meat control. The infringements of food control and price regulations by hitherto honest and conscientious butchers were more often than not dismissed with costs, and a warning under the Probation Act. However a repeat offence, or in cases of deliberate dishonesty, a custodial sentence or hefty fine or both were the usual outcome. Morally reprehensible, as all criminal activities are in war-time, it was very easy to innocently fall foul of the law with so many emergency regulations in place. With the exception of petty pilfering, organised crimes involving the meat trade were rare, although it has to be recognised that 'under the counter' dealing did take place. The black market existed in every sphere of society, from ladies nylons to petrol, but never seriously threatened meat supplies.

The most effective deterrent for minor breaches however was the glare of publicity by appearing in the local newspapers, the following offenders all butchers and their solicitors defending their actions, either true or false, 'A ticket had been moved by mistake from another tray'. 'I was short staffed at the time of the offence and the employee who normally priced the meat was on holiday'. 'I was standing in for my brother who has been called up'. Other examples of which in 1941 are the London Cooperative Society, Kentish Town Road and W. Bridger & Sons, Seaton Place, N W1, both pleading guilty to exposing meat for sale above the maximum price. In the same magistrates court Lidstone (Butchers) Ltd in Parkway were fined for a breach of the meat regulations and well known wallpaper manufacturers Shand Kydd in Highgate Road were similarly dealt with for keeping inaccurate records of canteen food. The last named business employing hundreds of people and by consequence a large provider of meals within a site area occupying 114,000 sq ft and several stories of floor space totalling 245,000 sq ft. The business was founded by Mr Norman Shand Kydd c1885 and incorporated in February 1918. In 1958 the company amalgamated with John Line, a leading manufacturer and merchandiser in wallpapers, paints and other decorators' supplies. A Royal warrant holder with retail showrooms nationwide, the most prestigious sited in Tottenham Court Road, London, with a pedigree no less impressive than Shand Kydd, having been founded in 1878 by the grandfather of the current chairman then being Mr J. B. Line. The company was incorporated in 1904 and had factory premises in Gordon House Road which undoubtedly influenced the twinning of the individual companies; thereafter known to city financiers under the umbrella title of K. L. Holdings Ltd.

Returning to shopkeepers, who themselves were often subject to criminality, the activities of one hapless burglar when the police followed a trail of turkey feathers leading from the shop to the culprit's home. At first denying he had a turkey in the house, stolen from Messrs Stevens and Steed provisions dealers 67a Camden High Street, another trail of feathers lead to a kitchen cupboard where the turkey was discovered, which was his undoing. The culprit was charged with breaking end entering the shop and stealing in all sixteen turkeys worth £25 pounds and receiving one of them. Although in Kentish Town high street lazy labourer John Smith aged 40 of no fixed abode found it more convenient to steal his Christmas dinner from the display outside the shop of James Bush, poulterer. In no particular hurry he calmly walked away and was immediately apprehended; his audacity rewarded with six weeks hard labour. The magistrate Mr Cooke commenting at his trial in December 1879, the local lads then employed to watch outside shop seemed to be as idle the thieves.

It was not all shortages and sing-a-longs in air raid shelters as the butchers trade, in common with other civilian volunteers, played their part in defence of the country at home or abroad. To begin with butchers over the age of thirty were included in an extensive list of reserved occupations, as an important element in the smooth running of the rationing scheme. The demand of the military for men however was eventually given priority and reserved status was withdrawn. The younger generation of butchers for the most part had already volunteered or been conscripted for military service. Many of them second or third generation sons of German migrants like Richard Bradley his father Cornelius Goebbels, master butcher at 94, Cambridge Road, Kilburn, in March 1936 changing the family name. His son Sgt R. Bradley awarded the military medal for bravery in the now legendary St Nazaire action and subsequent exploits escaping from a prisoner of war-camp. Others made the ultimate sacrifice; Lance-corporal Eric Harden a butcher in civilian life from Northfield, Kent and butcher turned soldier twenty-one year old Lance-Sergeant Jack Baskeyfield from Stoke-on-Trent, both awarded the Victoria Cross posthumously for gallantry.

For those left behind to keep shop, life could on occasions be similarly dangerous as most had joined one of the many essential war time services open to civilians. The first daylight air raids on London commencing the 24th August 1941, a precursor to the blitz proper found many of them on fire watching duties. The late Harry Harrison butcher at 257 High Street, Camden Town, situated between Woods greengrocers and Talby the fishmonger, recalled the lonely and terrifying experience of duty on the Royal Albert and the Victoria Docks. 'We used to call it pepper-pot alley' he said, luckily Harry lived to tell the tale.

As the war dragged on the hard pressed owners of butcher's shops were feeling the financial and practical strains of half empty fridges and uneven supplies. One butcher reported being delivered a months allocation of ox kidney in one week during an extremely warm period, another received only sweetbreads. Others in heavily bombed areas were experiencing a mass exodus of customers, while butchers in evacuation areas like Exeter city butchers for example with over 1,600 London and Hull resident evacuees reported trade buoyant. The London Butchers trade association complained of emergency feeding centres, renamed British restaurants on the insistence of Winston Churchill and one situated in Leighton Road, Kentish Town. That further exacerbated their plight, alleging

Figure 123: War Time Mobile Shop, Operated by R. Gunner Ltd.

not only were they loosing customers who were bombed out of their houses, but other members of the local population were going to these restaurants for food.

Another association representative asked why orders for supplying these centres were in the hands of certain traders, whilst others are having their trade depleted and suggesting a rota system. It was pointed out by the London County Council officials, the establishment for feeding centres were a definite policy of the Ministry of Food for feeding the public in times of emergency. Furthermore a rota system would entail a considerable amount of extra work and only a certain amount of meat was allowed to enter a district, so that the extension of meals served in the feeding centres would reduce the amount supplied to caterers through their respective butchers.

A similar feeling of discontent surfaced among the fifty-six butchers trading in Hampstead, when at the Chamber of Commerce meeting, Councillors F. Barrett, and R. Halse voiced disquiet concerning supplies of meat to A.R.P. canteens, asking why Hampstead butchers had not been given an opportunity to supply some of the meat. The outcome of the question a virtual restatement of that given to the London Butchers trade association was of particular significance to Frederick Barrett, as proprietor of a butchers shop in Englands Lane and other locations. The Hampstead butchers lost a stalwart member at this time with the death of John Draper in his seventy sixth year, who began trading in the 1880s from his shop at 11 Fleet Road.

An acknowledged expert on cattle breeding and a reputation on Smithfield market for being a sound judge of quality meat, he was eminently qualified to advise the Hampstead butchers retail buying committee. He was equally gifted in the world of entertainment which he demonstrated on numerous occasions, as a keen member of the Hampstead Strollers. It was here he entertained audiences with recitals and songs he had known from the days of the music hall, the work of Harry Randall in particular was a great favourite. In Kentish Town the number of retail butchers during the duration of the war appears to total thirty-eight a lower figure than in Hampstead which can be accounted for by the population of that area being widespread. Whereas in Kentish Town the high concentration of butchers within a compact and defined area would by implication mean limited opportunity for newcomers.

Irrespective of prevailing war conditions, the apparatus of a civilised democratic society had to be maintained. The Joint Industrial Council, following negotiations in 1942 on wages and conditions for employees in the retail trades, announced a new pay deal. The decision applied to those engaged in retail groceries, provisions, cooked meat and other cooked foods, fish, game, poultry, rabbits, fruit, vegetables, flowers, off-license trade and of course butchers. For meat trade staff located in London, the managers of shop branches wages increased with the shop turnover, but averaged from £3.15s to £4.15s a week. Shop assistants, van boys, cashiers, clerks and warehouse staff were paid according to age, boys at 16 received 23 shillings up to a maximum amount of £3. 8s aged 25 years, female staff of all grades received slightly less. The rates applied for a normal week of not less than 48 hours. The Minister of Labour and National Service pointed out that the agreement could not be statutorily enforced, but had the power to refer disputes to the National Arbitration Tribunal. He had in mind complaints from traders in defence areas, including towns where visitors are banned, making the higher wage rates unaffordable.

The war time reports of casualties and damage understandably were either vague or deliberately misleading, nevertheless many thousand of deaths, serious injury and damage were recorded. Among the tragedies, and one of the worst, was a direct hit on the Woolworth's and Co-operative stores in New Cross Road, the shops and pavements being busy with lunch time shoppers. It is estimated 1,200 shops in London suffered structural and stock damage. Although few butcher's shops were rendered unusable during the war, there was the problem of unexploded bombs and disruption to public transport and utility services requiring roads to be closed for lengthy periods. In areas where this proved a hindrance for customers, the few remaining mobile butcher's shops in London were pressed into war-time service and lorries were hastily converted to fill the gap by simply removing one side. The low incidents of severe damage to butchers shops are corroborated by a similarly low request for financial aid from the War Emergency Assistance Funds set-up by the local retail butchers buying committees.

The ration worries for the butchers' trade were not over yet, nor would they be for longer than anyone could possibly imagine. However as the war drew to its closing stages another difficulty began to emerge. The latest problem exercising their minds, and indeed the whole food sector, were the arrangements for food distribution on V.E. day (Victory in Europe), and the day following which had been declared a public holiday. It had already been made crystal clear by the government that no advance warning would be given, as the news of surrender could come suddenly at any time night or day. Without prior warning of such vital information there was concern that shops selling perishable food could close spontaneously as staff left to join in the celebrations, leaving customer requirements unsatisfied and supplies stranded in transit. The solution, in the case of the meat trade was far from straightforward, an announcement in the early part of the week, a traditionally quite period, would have minimal effect any later would have the complete opposite. All the Ministry of Food could do was recommend employers make known to staff and customers their intentions and should the news be announced during normal trading times, insure the shop remains open for at least one hour after. When the news of Victory in Europe finally came on the 7th May 1945, food shopping for a moment in time became the least consideration amid the celebrations.

Figure 124: St Pancras Town Hall, notice, December 1945.

MEATHIST MINIATURE

The term 'Corned' as in corned beef refers to the curing process whereby salt is an essential ingredient of brine, the word 'Corn' meaning a grain of salt, hence corned beef or corned leg of pork, the latter sold hot and eaten with pease pudding. The mottled appearance of corned beef is also associated with the slang term 'corned beef legs' a condition prevalent among women and attributed to poor circulation. A contributory factor often cited in days gone by, the habit if sitting too near open fires after exposure to cold.

15

TIME AND FASHION

After six weary years when personal triumph and tragedies, destruction and material shortages of every description had been a daily occurrence, the end of the war bought little respite to the long suffering civilian population, and the immediate post war period described by labour Prime Minister Clement Attlee, as 'winning the peace' and social historians as 'a war without bombs', was going to prove an almighty challenge. Within months of wars end the President of the United States, Harry S. Truman, announced the immediate ending of the Lend-Lease Bill and emissaries were hurriedly despatched across the Atlantic to negotiate billion dollar loans from both America and Canada. Britain was now mortgaged up to the hilt and facing serious financial and economic problems, which somehow had to be addressed by a nation already weary and exhausted.

For the Labour Government, swept to power on a tide of optimism and expectation, there emerged an unenviable dilemma of how to balance exports, imports, reconstruction and consumption against repayments of massive debts. Of these components, a competitive export trade at the expense of home consumption of food and supply of domestic goods were the only viable economic solution. The feelings of the British people were totally opposed to further sacrifices but under the circumstances there was no alternative. The relaxing of wartime food controls, except in the rarest of circumstances was completely out of the question; if anything they became progressively more stringent when in July 1946, bread the most basic of food commodities and never rationed even in the darkest days of the war was added to the list. The arrival a year later of the harshest winter since 1888 only exasperated matters still further, straining the already depleted coal stocks. In consequence electricity, gas and coal supplies to homes, shops, offices, factories and railways were drastically reduced or cut without warning to conserve coal stocks, at times immobilising entire areas of the country.

Desperate as the situation was it had not sapped our enthusiasm for business enterprise, on the contrary for the meat trade the road to regeneration had already begun when Josiah B. Swain, Master of the Worshipful Company of Butchers, undertook a tour of towns and villages, explaining and discussing its future plans. There followed an extensive good-will tour of the South Americas, where he met an array of prominent public and commercial dignitaries with whom he was able to lay the foundations for others to follow. Indication of the extent to which meat producing countries valued Britain as a trading partner were evident from the personal reception he received from the Presidents of Argentina, Brazil, Uruguay, and Vice President of Chile. In Buenos Aires he entertained President Peron and Senora de Peron to dinner at the Plaza Hotel. There were other good-will visits to America in 1948 to the Master Butchers Association, all in preparation for a time when meat control would be abolished and the full potential of overseas, and crucially our home meat markets could be realised. The job in hand for retail butchers was to plan for the future when they could modernise and expand. Waiting in the wings ready with expansion plans of their own however were the grocery businessmen, back from America with ideas and innovations far and above any conceived by our large meat multiples.

The return to supposedly happier days was marred by frequent local and national strikes by workers impatient for change. In 1947 the army was called in to distribute meat from Smithfield market to London retailers, The disruption played havoc with shop supplies and the butchers received the blame from uniformed grumblers and journalist, who received among others swift retaliation from the pen of Leonard Harry Kimber, Chairman of St Pancras Butcher's Buying Committee and son of G. F. Kimber a Kentish Town retail butcher. Yet little by little there would be improvements along the way; in March 1949 when clothes rationing, but not price regulation ended and the sweet tooth of the nation pacified when sweets were taken off ration in 1949, however such was the stampede suppliers could not cope and restrictions were re-imposed for a short period.

In between the set backs and small successes emerged a situation with the potential of inflicting a damaging psychological blow to the nation when Britain hosted the 1948 London Olympic games; the opposite of intention that in essence was to lift the spirit of a tired nation and raise international prestige. The so called Ration Book Olympics opened a national debate beginning with the question; how will the competitors and officials be fed amid the continuing food shortages. The Government, despite statements to contrary resorted to subterfuge and the athletes were granted category 'A' status meal allowances applied to heavy manual workers such as dockers and coal miners in addition supplemented with two pints of liquid milk per head per day and half a pound of chocolates and sweets per head per week.

To ease the meat burden generally among fellow athletes the Argentines bought with them 100 tons of meat and Iceland offered frozen mutton, other competing nations such as Holland offered fruit and vegetables. Food supplies at one of the Olympic housing centres in Uxbridge were said to be ample thanks to the generosity of the American contingent who arranged daily deliveries of fortified white flour aboard charted flights from Los Angeles. The deliveries causing the engaging if mildly eccentric future television personality Dr Magnus Pike then employed in the Ministry of Food, Catering Division to express the opinion the Americans as having a "disregard for geography and expense". The understanding between the competing nations forewarned of food shortages and the London Olympic Committee trying to address the situation did not last long as quantity and variety of food on offer failed to satisfy many of the athletic teams or their representatives who managed to complain almost on a daily basis. The games although undoubtedly successful, the Minister of Food, John Strachey in order to pacify an anxious public; stated the estimated increased food consumption of participating national teams and officials during the Olympic Games amounted to 0.16 per cent "a completely insignificant amount".

In 1950 bread, eggs, flour and soap were de-rationed, and the government even felt confident enough to dispense with publishing the irritating food facts leaflets. All of which reads like the hard times were practically over, however nothing could be further from the truth. In reality, when faced with butchers still obliged to sell imported tins of veal at 2s.7d and mince at 1s .4d on the points system, to augment the meagre fresh meat ration the housewife queued in solemn fatalism that rationing was never going to end. On 27th January 1951 the meat ration was reduced to 8d, meaning a drop in profits for butchers while overheads continued to increase despite government rebates to ease the financial burden. The meat trade reacted vehemently with the London Butchers Association circulating within its membership a call to dress their windows as in mourning, with black crepe paper, for a nation now fed on eight pennyworth of carcass meat and two pennyworths of corn beef. In one satirical cartoon, butchers were advised to exchange their cleaver and knife for a magnifying glass, calliper and razor. The National Meat Traders Association, in hope rather than anticipation of any improvement, implored their members not to be disheartened, but to keep abreast of technology and education in the trade.

In anticipation of rumours circulating in 1954, hopefully true this time, that meat would be un-rationed by the end of the year, or D-Day as the trade called the approaching decontrol day. In preparation the doyen of the meat trade Frank Gerrard was invited to appear on "About the Home", a BBC television programme for housewives transmitted fortnightly on Thursday afternoons. The first of a series of programmes begun in January, in which he explained to viewers the various joints and relation to the carcase, while cookery expert Marguerite Pattern showed how to make use of the less familiar cuts in the home kitchen. As meat was still on ration, the reaction of viewers fortunate enough to have televisions I shall leave to the reader's imagination. One irate butcher wanted to know did the Ministry Of Food supply the Scotch beef, as he had had nothing but cow beef for months. The question of who supplied the beef soon became irrelevant, when a few months later after nearly fourteen years the Food Minister Major Lloyd-George speaking in the House of Commons announced 'it is the Governments intention that rationing, price control and distribution of meat and bacon by the Ministry will end at midnight on Saturday 3rd of July 1954.'

Figure 125: About the Home, BBC T V 1954.

To draw a line under this long and difficult period, it can be said the immediate post-war period will be remembered as a time when nature, man and misfortune conspired to create a period of sustained national depression unequalled in the latter half of the 20th century. As to what the ration book represented, some women grew fond of them, some cursed them and a few died for them, but all in all they served the nation well. The butchers gave a sigh of relief and festooned their shops with 'Anyone Served' notices. Although for office workers, it seemed they were destined never to be free of some kind of official authorisation to eat. For in 1955, on the first day of the first month of the first full year without ration books, Luncheon Vouchers Ltd began operations. The vouchers issued in dominations from 1s to 5s each were used primarily by office companies as an extra incentive to employees who exchanged them for meals.

Figure 126: Sainsburys, Fresh Meat Counter Service, 1956.

OLD HABITS

To say little had outwardly changed in the butchers shops of Kentish Town, as elsewhere since the end of the war, would be an understatement. Apart from the freedom of choice, there were no apparent earth shattering changes within the retail trade. The butchers settled back into their cosy pre-war shop routine, and Mrs Post-rationing came in on the same old days and was served in the same old way. The butchers were about to get a wake-up-call from a changing retail world that would eventually leave many struggling to survive. At the forefront, and revolutionary at the time would, be the emergence of self-service stores, now universally known as supermarkets that heralded a new era in High Street shopping. The seeds of self-service stores had been sown in America; way back in 1912 and within twenty years had mushroomed into thousands of such stores across the country. On this side of the Atlantic, planning by leading grocery companies was already well advanced to introduce a similar shopping experience once post-war conditions allowed. In reply to the often asked question, who was the first to use the word supermarket on their store in this country, we can only answer with the saying 'success has many fathers, but failure is an orphan'. From a slow start the entry of supermarkets into the retail market place gathered pace in the second half of the 1950s, and within a decade these so called temples to consumerism were an established retailing fact.

In Kentish Town we have a well documented account of this transition in 1954, when J. Sainsbury's decided to open the first self-service store in the area and within eighteen months a sparkling new glass fronted store had been built in the High Street. To minimise disruption of sales, the old branch shops had been scheduled for closure on completion of the new store sited at 250-254 Kentish Town Road, a few hundred yards away on the corner with Islip Street

The changeover on Monday 5th December 1955 during the Christmas trading period required a high degree of military precision. In the old shop daily routines were observed and windows were dressed with seasonal items and customers placed their orders as the build up to Christmas carried on as normal. On the final day, two equally important people but with different responsibilities, were in the store. Long serving employee Herbert Pither, poultry man who started in 1919 as a delivery boy with the company, and Alan Sainsbury, grandson of the founder who among others pioneered the concept of self-service supermarkets in this country. Both men aware a new era was beginning as the shutters came down on the old branch shop for the last time and next morning at eight o'clock the new self-service store opened for business.

The meat department was divided into two sections with shop floor sales area comprising of refrigerated wall cabinets parallel with counter service while the floors above housed the meat preparation, wrapping and weighing area with communication between sections by means of microphone and speakers. A lift was provided for heavy goods, or alternatively a chute. All butchery staff wore white protective clothing although job titles differed from those in traditional butcher's shops using titles of meat supervisor, head and assistant head butcher etc. At the end of the first week of trading, the five check-outs were declared by Alan Sainsbury to have been 'satisfactorily busy since opening day'. There were to be many more satisfactory weeks in the years ahead until the store closed, together with a shop at 151 Queens Crescent, on 2nd November 1968 to be replaced and upgrade by Sainsbury's new supermarket opened on the western side at 217-223 Kentish Town Road. On the 15th April 1989 the unthinkable happened, when J. Sainsbury severed its seventy three year history with Kentish Town Road, by closing the supermarket and relocating to Camden Road. Of passing interest, untypical of a major high class grocery business, the company it seems never sought or applied for a Royal Warrant; despite the fact their arch rival Lipton Ltd. displayed the royal coat of arms above their shop at 161 Queens Crescent adjoining Sainsburys.

Figure 127: Sainsburys, Self Service Fresh Meat Cabinet, 1956.

In the Camden Town area another early exponent of self-service was Anthony Jackson, located in Inverness Street, this was replaced by Downsway Supermarket Ltd. that had a meat department and was sister company to J.H.Dewhurst.Ltd.

The butchers' reaction to supermarket encroachment into the traditional butchers trade was viewed with indifference 'A five minute wonder', a short sighted view that mirrored the introduction of frozen meat some three quarters of a century earlier. Others doubted the necessary butchery skills available among supermarket staff; especially the traditional trained butchers oblivious to the fact many of their number had already deserted the austere working conditions that existed in butchers shops for the congenial environment of supermarket life. We must not blame the butchers entirely, proprietors or employees, as many of the problems were outside their control; their decline in our communities was not one gigantic explosion of supermarkets but a slow process with many other causes. At the onset of the supermarket revolution butchers shops numbered 45,000 nationwide, it was unthinkable they could ever be replaced. The very idea the meat eating public would even contemplate buying meat in cellophane wrapped cardboard boxes was greeted with ridicule and complacency. They had much to be complacent about in those times, what the customer could not see, touch or smell they distrusted. A long running prejudice against the contents of tinned meat fostered by their forebears and reinforced by war time rationing had taught them so. While all around them adapted to a changing lifestyle of shopping, the independent butchers twiddled their thumbs like King Canute challenging the tide of modern consumerism to advance.

There misguided attitude bolstered by the lure of free enterprise and being ones own boss, that for a while in the 1960s ensured a continual stream of new recruits eager to replace the astute proprietors that decided now was the optimum time to vacate their shops. Those that began there reign in Kentish Town and beyond at this period included Corrigan, Gilry, Hyde, Drake, R & C, Daniels, P & G, and L. E. Tucker and many more. The last named butcher situated at 81 Highgate Road and frequented by the distinguished Australian journalist Peter Smark (1936-2000). The shop one of several aspects of English life in company with Parliament Hill Fields, Hampstead and the madness of Camden Council lamented by the former foreign correspondent.

CHANGING LIFESTYLE

By nineteen-seventy there were already signs of staffing problems due to full employment and the need for large capitol investments on shop modernisation that would make many small and medium sized companies begin to question the financial viability of such retail operations. The mobility of affluence and absence of Mrs Supermarket was beginning to impact on the meat trade, and long established butchers shops began to disappear from our towns and villages. The days when generations of customers were served from pram to retirement age were dwindling fast. The trade was thrown a life line during the nineteen seventies with the home freezer boom, when whole carcasses of beef, lamb and pork were bought and pre-packed for the months ahead. A short lived experience, as housewives soon discovered that without a strict menu regime the choicest cuts would inevitably be eaten first leaving a diet of the less appetizing meat based meals. Even less appealing the growing number of retail companies who sold home freezers on condition the purchaser also entered into an agreement to buy meat supplied by the same company. The interest charges in many cases exorbitant and the meat often of poor quality sold at inflated prices and on closer inspection a considerable number of agreements found to be totally illegal.

Figure 128: R. Tuckers, 177 High Street, C. T. 1975.

Fortunately for Kentish Town customers that had been duped a Consumer Clinic opened in 1965 by the Consumers Association, publishers of Which magazine in response to the consumer boom was on hand to provide a much needed source of consumer protection. Later Consumer Aid Centres began to appear nationwide, one partnering the Citizens Advice Centre at 242 Kentish Town Road.

Of major significance in this decade was the conversation to decimal currency from pounds, shillings and pence; a currency fundamental to the British way of life with a long and eventful history, that numismatists generally agreed had begun in the reign of King Offa of Mercia (757-96). The build up to 15th February 1970, the official day when the country changed over to decimal currency, began several years before; with coins being stockpiled for gradual release within the current monetary system. There was a foretaste in 1968 of what might lay ahead, when several incidents occurred; bank tellers accidentally issued batches of the new decimal five and ten pence pieces equalling the old one and two shilling coins ahead of the official release date. One tradesman, a Birmingham butcher, initially refused to accept them until a Treasury spokesman assured him and other concerned shopkeepers they were in fact legal tender and the first such decimal coins to be issued in the United Kingdom. When the official 'D' day (decimal currency day) arrived the practicalities of transition were less traumatic than the retail trade and general public had expected. Although it was all too much for 80 year old master butcher Tom Tuckey in Warwickshire, who decide to close his shop and retire.

From the nineteen eighties the downward spiral of closing butchers shops gathered pace and continued throughout the nineties and into the new millennium. In addition to those already mentioned we may add the Camden Town shop of Edward Harrison sited in a parade of high street shops, many of them butchers, between Inverness Street and Jamestown Road. This business was a family concern with Edward at the helm controlling another two shops in Bethnal Green Road, East Ham. His Camden Town branch adjacent to Woods the greengrocers was managed by his brother Harry, and opened prior to the Second World War until closure in the nineteen seventies. A rather small but elegant looking shop with bronze window surround and a decorated tile and marble interior, together with stainless steel wall and window hanging rails; the white embossed ceiling complementing the mahogany cash office. The entrance with shop title above of the roller shutter kind and the only way into the premises', the rooms above during the Harrison tenure remained unoccupied and were used for storing carrier bags, wrapping paper and laundry. The basement consisted of a labyrinth of steps and preparations rooms, which also housed a baffle board refrigerator. A covered passageway led to a small sheltered area with gas copper and into the backyard where a freezer had been installed. Overlooking the backyard a L.C.C. depot used for storing building materials that unintentionally housed a population of feral cats and directly behind towered Rowton House, Arlington Road opened in 1905 and named after philanthropist Lord Rowton dominated the skyline. The Harrison building at number 257 now one of historical interest as in a room above Tom Sawyer (1826-1865) the celebrated pugilist died here of tuberculosis. A blue heritage plaque placed above the second floor windows in 2002 commemorates his short life. He died on the 8th November aged 39 years old and is buried in Highgate Cemetery. His grave instantly identified by the presence in stone stature of his pet bull mastiff exemplifying his master's tenacity and courage.

Figure 129: Hampstead Butchers, 17 High Street, 2005.

By the mid 1980s a host of retail butchery businesses both independent and multiple were reporting adverse trading conditions, and eventually ceased trading, others moved into wholesale supply and other commodities especially poultry. Others engaged on a policy of corporate expansion and became subject to mergers and take-overs. The family based independent butchers were only marginal better off and many closed for want of the next generation taking over the business, instead their sons and heirs preferring the security of long term financial investment in property rather than the uncertainty and daily hardships of shop life. Those remaining in trade found recruitment of staff difficult, for with the exception of supermarket butchers the remuneration for shop butchers were often less than a factory worker. In addition the majority of independent and multiple butchery concerns were still operating under the one or two half day closing regime; this condition of employment wherein employees rarely started their half day before 2 pm and not then if awaiting a late delivery; later generations of prospective recruits found unacceptable. A reminder of the average weekly wage rates of £47 for butchers and £32 cashier/counter hand in Robert Tuckers advertisement for staff. He also traded for a time at 21 Theobalds Road, opposite the Camden Local Studies and Archive Centre, the business was dissolved in January 1997.

Figure 130: Barnard Butchers 86 Plender Street, Camden Town, 2003.

A changing life style of takeaway meals and supermarket shopping created an unstoppable decline that even the medium and large multiple butchery company's could not survive. That ushered in public notifications of branch closures to customers and redundancy letters to employees, as companies contracted their operations. One such public letter from the head office of Kingston Brothers (Butchers) Ltd., to their branch, opened in 1937 at 102 Mill Lane, West Hampstead, explains in detail the reason. On this occasion staff was transferred to another branch in West End lane as a temporary solution in the forlorn hope trading conditions would improve. The final outcome was one of inevitable closure against the advancing force of social change and technology.

Other independent shops including butchers fell victim to the domino effect, whereby one by one traders closed leaving a particular business isolated, conversely the closure of a single essential service, most notably the post office that deprives other traders of customers. The juxtaposition of efficiently managed and financially profitable shops finding themselves in changing circumstance whereby the culminating loss of customers leads to an unsustainable reduction in profits.

Among the myriad of reasons given for closure have been property values, exceeding business potential, retirement, and widespread adverse publicity following food scares effecting consumer confidence and the cost and time to implement HACCP (Hazard Analysis Critical Control Programme) a systematic preventive approach to food safety. However time has evaporated the argument against the ever present supermarket and is unsustainable in today's society. Competition between trades is not a modern day phenomenon as the following extract taken from the St Pancras Chronicle demonstrates. 'In the evening of the 5th of December 1941, Councillor P. G. Owen Chairman of Hampstead Chambers of Commerce gave an address titled "End of Hampstead Small Trader?" in the King of Bohemia public house. The nub of his argument being that local traders must rally against the demise of local shops caused as a direct result of crippling legislation and the competition of multiple chain stores'. Since and before that time there have been many such debates and rallying calls to save the livelihood of businesses within the community. In the opening years of the new millennium there are signs that the national descent of butchers shops into obscurity is levelling out, albeit closures are still occurring.

In November 2002 Barnard Butchers in Plender Street, Camden Town pulled down the shutters for the final time after fifty years of trading and the peeling paintwork, broken weather blind and builder's paraphernalia inside 'Hampstead Butchers' (1991-2005) tells its own story. In the photograph the embossed plague on the brickwork dated 1888, just visible to the left above the shop facia commemorates the development of this part of Heath Street joining with Fitzjohn's Avenue. The fabric of the shop front had nothing to commend it for posterity unlike the similarly named and now defunct Hampstead Quality Butchers near the Hampstead bus terminus, South End Road/ Green. The shop frontage from an earlier period has been preserved whilst from the nineteen-nineties retaining the butchers business title above and is currently trading as Polly's café.

Figure 131: J. A. Steele, 8 Flask Walk, Hampstead, 2005.

On the periphery of Hampstead Heath is Flask Walk, often over looked by the casual visitor, this side street still retains much of its charm and Dickensian feel. The once narrow entrance until 1911 with over-span and windows of the bridge rooms above sharing space with advertisements on behalf of Riding Master F. Davy and Hovis bread. In the 1870-80s the services of butchers Joseph Revitt, Joseph Harrison: a baker, fish and cheesemonger, tobacconist, grocer, boot maker and hairdresser were available within the short walk to arrive at the Flask Tavern.

Figure 132: Highgate Butchers, 76 High Street, Highgate, 2005.

The fishmonger later included the fried variety with chips, this traditional of English fair prepared by Mrs Anne Taylor, well known proprietor in the 1950s. Other notable retail butchers in this road and adjacent High Street were Druce & Craddock, F. Alexander, F. Barrett, M. Head, Lidstone Ltd, W. O'Hara and Swatland & Sons.

The resident butcher in Flask Walk since the mid 1960s had been Joe Steele, serving notables from the world of film, theatre and literature amongst his cliental. In realisation of the trend towards greater culinary expectations he published the 'Hampstead Cookbook', a collection of meat, poultry and game recipes, many contributed by customers. After a lifetime of service spent in the butchery trade, seventy-three year Joe decided to retire and in December 2007 the shop closed. In January the following year the building was acquired by CP Plus, a car park management firm who occupied the adjacent premises with the intention after refurbishment of re-letting the property to another food retailer. To the south in Camden was Franks Butchers at 188 Eversholt Street, established in 1990 and recent casualty (2010) of a declining trade.

THE JOLLY BUTCHER

So what has become of our local 'Jolly Butcher' with his straw hat, stripped apron and large corporation, a ready smile and a 'good morning madam'? He is still there but today you will have to seek him out, albeit the fellow with straw hat and rotund physic so beloved by caricaturists is a very different character today; replaced by health conscious forward looking highly professional craftsman, ready to meet the expectations of each new generation of consumers whilst still preserving the best traditions of the trade.

In 1894 the following appeared in a national newspaper 'The trade of butcher is not one calculated to develop delicacy of feeling in those that follow it, rather the reverse'. Just over a century later in 1999, concern was expressed in meat trade circles that as an industry there were far too many graduates in the trade and not enough retail butchers.

The clock of history cannot be turned back for the local butcher, the hey-day of the butchers shop belonged to our parents, grandparents and great-grandparents and if a certain age in our own time.

A constant theme throughout this book has been one of change; no doubt previous generations regretted the passing of the milk-maid just as we regret the passing of so many local high street shops. Fads, fashion and time decree everything has its day in the sun and change is not necessarily a bad thing if replaced with something better. Although attention must be drawn here to an earlier period of consumer history, wherein economic and social change were also responsible for the gradual demise of one shopping experience and the creation of another; for it was shops in the 18th century that began to replace markets and fairs as the principle places to purchase all kinds of commodities. As today, disgruntled sections of the public bemoaned their passing. The situation then was somewhat different in that markets were not always a permanent feature of a close knit community.

Figure 133: Barretts, Englands Lane, Hampstead, 2005.

However let us not delude ourselves, for when we abandon the modern conveniences of a purpose built air conditioned shopping mall to visit a street or country market, we fulfil a psychological need within us to forage for our food. Nevertheless their role as providers of the necessities of life was equally important. In brevity, that like today only where and how we buy our food has changed not the availability. In consequence only a few retail butchers in the Borough of Camden created in 1965 from the former Metropolitan Boroughs of Hampstead, St Pancras and Holborn have survived. Some of whom and ignoring the anomalies of shifting boundaries are discussed in these remaining pages.

In the picturesque area of Highgate mother and son partnership, Phyllis and Lee Harper of Highgate Butchers in the High Street celebrating thirty years of trading continue the tradition of retail butchers in this road. The shop at periods in history administered by Hornsey Local Board and after within the Borough of Hornsey and now officially in the Borough of Haringey. Mrs Harper herself joins a select band of women butcher proprietors in the meat trade, not least Mrs Elizabeth Attkins who traded from number fifty-five in this high street over one hundred years ago; both widows of master butchers who bravely continued in business, despite the great loss of their husbands. A trade of long hours and physical demanding work behind the scenes; with as much work to be carried out at the end of the day when energy is at low ebb as there is at the beginning of a days trading.

Often not realised, an important part of the skill required by a retail butcher is the ability to disassemble and prepare an article for sale, distinct from the majority of trades that assemble an article from ready made components, a skill which is almost unique in producing a natural food product for sale.

The traditional butchers Barretts in Englands Lane, has since the early 1980s been in the capable hands of Bob Enright the proprietor. With justifiable pride he remembers starting as manager at the shop forty-two years ago. A craftsman butcher of the old school, while acutely sensitive to requirements of modern day consumers, he continues a valuable service at this location stretching back over a century. The left hand side of this road beginning at the Washington Hotel where most of the shops were situated, including the Victoria Wine Co, was originally named Elizabeth Terrace. The title Englands Lane originally began on the right hand side at St Mary's Convent interrupted by Chalcot Gardens before continuing at Wychombe Studios, then home to artists Edgar Barclay and Alexander Burr. The rows of shop fronts, with decorated trusses, pilasters and iron balustrades above, still much in evidence today, alas missing are the ornate late Victorian street lamps from a bygone age then in place, the entire length of the road lending an air of opulence to the lowliest of trades.

Others survivors still trading along include Theobalds butchers, sited opposite the Camden Local Studies & Archive Centre in Theobalds Road. The shop as with butcher James Grubb at 85a Leather Lane, formerly within the Borough of Holborn, which between 1900 and 1965 was the smallest of the 99 Metropolitan Boroughs and also one of the wealthiest. The business in the hands of McKenna Meats Ltd a combination of catering, wholesale and retail meats. Among previous proprietors Henry Collingwood of Great College Street Camden Town 1868; a family with generations of meat trade history and from a later period Gregory's butchers 1969 and R. Tucker (Butchers) Ltd, 1994. In Camden Town Miguel Martins of Martins Butchers has run the former Tucker branch shop in the High Street since 1992.

Figure 134: Theobalds Butchers, 21 Theobalds Road, Holborn, 2007.

In Kentish Town from approximately 402 retail butchers shops that have opened for business during its history, only three are trading at this point in time of these one sited away from the high street and occupying a corner position where York Rise intersects with Chetwynd Road is the butcher's shop owned since 1989 by Maou Ghosseiri who still uses the title Jackson Bros, the name of the previous occupants. It is a friendly, clean and airy shop and readily accessible to nearby residents. A peculiarity perhaps of times past on the top right hand corner above the shop front can be seen the road name plate York Rise although the shop address is as stated in the picture caption. Which may indicate the building was either accessed, or had an additional window facing Chetwynd Road. As with the majority of retail businesses a steel shutter is lowered at night as a necessary precaution against vandals and justified by the graffiti that appears next morning. An iron curtain where once the shop keeper dressed his window for the hours of closure with all manner of items particular to his trade for the interest and sometimes amusement of passers by is almost at an end. Although there is no doubt as to which local residents would rather see and shop proprietors would prefer.

Figure 135: Jackson Bros, 69 Chetwynd Road, 2003.

Further north off Highgate Road are Elite Meats in Swains Lane, (est. 1986) owned by the affable Martin Leahy an apprenticed served butcher with a life time of experience in the trade. A local born man, both he and his shop epitomise all that is best in a neighbourhood retailer. The premises situated in a pleasant tree lined paraded of utility shops, enhanced by an imposing Tudor style facade. The spacious shop interior completely refurbished during the nineteen thirties still retains the original wall to ceiling white glazed tiles complemented by a green and brown dado, reminiscent of the art deco period. It was during the incumbency of the previous butcher Silvanus Webber this work was undertaken. Although the interior shop furniture and equipment has long since been discarded for its modern day and hygienic equivalent. The firm was a limited company for forty-five years and better known to earlier generations of shoppers with branches throughout Camden. The business today specialises in English and Continental prepared meat, reflecting the diverse culinary preferences of local shoppers. Swains Lane in bygone days was known as Swine's

Lane and marks the northern limit of Kentish Town. Then a narrow road between houses, the entrance was widened in 1894 when a piece of land forming part of the garden of the Vicarage of St Anne's Brookfield was purchased by the Vestry for £442 19s 6d, the approximate cost today £25,000 pounds.

Figure 136: Elite Meats, 21 Swains Lane, Highgate Road, 2003.

To complete this brief look at some of the butchers shops currently trading in the Borough of Camden is The Pure Meat Co & B & M Seafoods butcher, fishmonger and poulterer adjacent to the Oxford Tavern public house in Kentish Town Road. The owner is Iranian born Ahad Dasht the son of a fisherman who prefers his customers to call him Harry in keeping with the friendly atmosphere of the shop. The only accredited organic meat butcher and fishmonger shop in the area which also counts well known celebrities among the clients, although whoever shops here is treated with courtesy from Harry and staff. . The company recently began a new venture nearby with the opening of Café Red that encompasses the whole rage of organic food and including a jazz club. Prior to his arrival in 2002, the business was in the hands of Barbra Burchell and Phil Fairman the latter of whom it was said extolled patience, kindness and a high degree of butchery skill; making his death in July 1992 at the early age of 38 all the more regrettable.

Figure 137: The Pure Meat Co, 258 Kentish Town Road, 2005.

Traditionalists would question the validity of a fishmonger dealing in butcher's meat. Yet animal flesh from whatever animal species is considered by many to be meat. A broad definition contrary to the following familiar request: 'I won't have any meat this week butcher; 'I will have a chicken instead'. A pedantic viewpoint and somewhat old fashioned in today's fiercely competitive world of food retailing where a measure of diversity is essential if retail butchers are to survive. The anomalies of classification presenting different view point from a legal perspective. In 1952 the Lord Chief

Justices Parker and Jones had before them the vexed question whether fish were included in the definition of "meat" in section 125 of the Transport Act 1947. On this occasion and for this purpose it was deemed the word meat did not include fish. Although medieval history also records it was once common to speak of "meat" as meaning "food" the usage of the word still in evidence in the days of Dr Johnson.

Of specialist interest there has also arisen through the ages differing interpretations and confusion surrounding the word 'dressing' associated with the preparation of poultry. By common familiarity the word has become an all embracing description for several distinct functions: Dressing poultry is de-feathering (plucking) and singeing, Drawing is removal of the internal organs (evisceration) and removal of head and feet and Trussing is fixing the legs and wings with string or wood faggots (skewers) ready for cooking. The term Blind Trussing is where neither is used, but the same effect is attained by manipulating the wings and legs.

The Pure Meat Co shop has filled a variety of roles during its history from coffee rooms to confectioner, until the eighteen-nineties and thereafter becoming a fishmonger's. The first and suitably named occupants in this trade Alfred and Elizabeth Starling carried on later by their son William, before passing onto the Carter family with a long history as fishmongers in this road. A brief mention of Fish 'N Fowl, 145 Highgate Road trading in the millennium year 2000, that while unsuccessful at this location the proprietors Robert Ifrah and Adrian Rudolf are worthy of praise for their attempt to revive the art of the fishmonger.

In the year 2000 the Food Standards Agency (FSA) imposed a compulsory system, whereby retail butchers shops dealing in raw meat and ready to eat food were required to be licensed and subject to local authority inspection. An ill thought out and cumbersome piece of legislation opposed by the butchers as unjust and unnecessary because restaurants and other such food establishments were exempt and unnecessary because stringent food compliance regulations were already in place and applied equally. The mandatory licensing regulation for butcher shops as predicted in the previous edition of this book was revoked on the 1st January 2006. The number of licensed retail butchers shops in the Borough of Camden in 2005 stood at 13 of which 5 of these were Halal butchers. The approximate numbers of retail butchers in the United Kingdom as calculated by the Meat and Livestock Commission in 2002 was 8,500, of these 500 are in Wales, 800 in Scotland and the balance in England, which include 400 Halal and Kosher outlets..

There are additional outlets selling meat, such as local franchise stores, garages and others that are not included in these figures. The overall number today is between 5,000 and 6,000 still trading wherein the average age of butchers is fifty years old. Of these, large numbers have diversified into kitchen ready, bakery products and non food items like barbeque cooking equipment where meat sales only account for a small part of the business. Those with larger premises providing a seating area for coffee, tea and snacks, in some instances restaurants wherein the customer may select their meat dish from a butchers counter. One enterprising butcher prepared takeaway curries and other meat based meals during the day for sale in the evening through a closed off section of the shop, another in the dual role of village butcher and post office. To survive others have adapted sections of the shop to incorporate gourmet foods, wines, exotic fruits, and vegetables.

The craft of small goods production in the retail butchers shops has for the most part been replaced by Kitchen Ready meat products that are bought in from specialist suppliers in pre-processed form requiring only minimal preparation, or obtained ready for sale from meat product wholesalers. In contrast, high class restaurants and hallowed food halls of our premier stores have reported a dramatic increase in sales of offal, pigs trotters and calves feet etc., once considered only fit for the poor mans restaurant. They are now much sought after delicacies; one mans necessities has indeed become another's luxuries. In assessing the role of supermarkets in the decline of retail butchers shops, it would be wrong to presuppose that in every circumstance the public preferred to choose their meat requirements from a self service cabinet because on family celebrations days, public holidays and other festive occasions, especially at Christmas considerable numbers of the public return to the traditional butcher.

The reign of the multiple retail butcher is over; there will be no return to the halcyon days with a branch in almost every high street, unless from public demand. That era has passed into history along with the self made men who created them. Although it maybe said the remaining independent butchers have enjoyed something of a revival in popularity during the last few years as shoppers have become increasingly aware of health issues and assurances over traceability and quality of meat, bringing a return to shop titles with generic animal names prefixed with Ginger Pig, Pot Bellied Pig and Lazy Lamb. Of the long term future of retail butchers shops we have no means of knowing. Will we ultimately receive our Sunday roast by post courtesy of on-line computer shopping or the return of the butcher's boy carrier cycle synonymous with the traditional trade? In 1866 the United Albion Circular, forerunner of the Camden & Kentish Town Gazette, printed on the front page 'The butcher's shops are the pride of old England' a view that still resonates with many people today.

MEATHIST MINIATURE

The widely used cellophane for product packaging was invented by Swiss chemist Dr. Jacques Brandenburgher in 1908 and first used on cigarette packets by Carreras on the Craven 'A' brand. The Carreras Arcadia works and offices (Black Cat Factory) opened in 1928 and were sited in Mornington Crescent, Camden Town. The further development of cellophane as PVC, (polyvinylchloride) revolutionized the food industry.

RESEARCH

The reader will be surprised to discover that, despite our reputation as a nation of shopkeepers, there is scarcity of books and material dealing exclusively with the development of the retail butchers trade. On the subject of meat trade books, these fall into three categories: primary history, technical and privately published family business histories. Of these only a small minority of the retail multiple and independent butchers published a business/company history, so the researcher must be prepared to collect material from widely scattered sources.

1) Camden Local Studies and Archive Centre Holborn Library, 32-38 Theobalds Road London WC1X 8PA.
   Here you will receive advice and assistance in finding material from dedicated staff.
   Website: www.camden.gov.uk/localstudies

2) The Public Record Office, Ruskin Avenue, Kew, Richmond, Surrey TW9 4DU.
   Holds a selection of business class records of dissolved Limited Companies, dating from 1844 to 1973.
   Website: www.pro.gov.uk

3) Guildhall Library Aldermanbury, London EC2P 2EJ. Manuscript Section:
   A collection of various records of The Worshipful Company of Butchers, dating from 1658 to 1950.

4) National Register of Archives:
   This is a search web-site service with names of meat traders and who holds information on them.

5) British Library, National Sound Archives:
   This archive has recordings of oral history contributed by retail butchers and others in the meat trade.

6) The British Library Newspaper Library, Colindale Avenue, London NW9 5HE.
   This library holds copies of the Meat Trade Journal & Cattle Salesman's Gazette, dating from 1888, and other trade periodicals.

*All the above have Web Sites

BOOK SOURCES

The Builder Magazine
Business Archives Council
Camden New Journal
Camden History Reviews
Cramer, Philip, Private Papers
Dewhurst (Butchers) Ltd.
Goad Insurance Maps
Hampstead & Highgate Express
Katz, Michael, Private Papers
London Gazette
London Metropolitan Archives
Meat & Poultry News
Meat Trade Journal & Cattle Salesman's Gazette
Newsletter & Bulletin, Worshipful Company of Butchers
Old Bailey Records, London. 1674 to 1913
Old and New Hampstead Illustrated (c1893), Unknown Author
Public Record Office
J. Sainsbury plc. Archives
St Pancras Rotary Club
St Pancras Vestry Minutes
St Pancras Parish, (1856-1901), MOH Reports,
St Pancras, Rate books
Sun Fire Office Registers 1816-1824
Times Newspaper
Trade & Street Directories
City of Westminster, Archive Centre.
Various: Correspondence, Legal Documents, Manuscripts, Memoranda, and authors collection.

FURTHER READING [SELECTIVE LIST]

Bonser, J. K.: The Drovers (1970)
Booth, Charles: Survey of London
Camden History Society: series Streets of, Kentish Town, Gospel Oak, Highgate (2005/6/7)
Corsair B.A. & Fitzell W.L.: 'The York Butchers Guild', (1975)
Crilly, Marlene: Blitz Kids (1990)
Critchell, J. & Raymond, J.: A History of the Frozen Meat Trade (1912)
Dunning, James: Britain's Butchers (1985)
Gibbons, Sue: German Pork Butchers in Britain (2001)
Jones, Phillip E.: The Butchers of London, WCB, Guild History (1976)
Keevil, Ambrose: The Story of Fitch Lovell (1972)
Mayhew, Henry: London Labour and the London Poor (1856)
Middleton, John: View of Agriculture of Middlesex (1800)
Nelson, A. A.: N.F.M.T. Pork & Bacon Section 1917-55, (1955)
Nicholas Courtney: A Cut above the Rest, WCB, Guild History (2005)
Pawley, Margret: Servant of Christ (1987)
Perren, Richard: The Meat Trade in Britain 1840-1914 (1978)
Richardson, John: Kentish Town Past (1997)
Rixson, Derrick: The History of Meat Trading (2001)
Tindall, Gillian: The Fields Beneath (1977)
Trow-Smith, R.: Livestock Husbandry To 1700 (1957)
Wilson, Charles: The History of Unilever (1954) Vol 1& 2. Unilever 1945 to 1965 (1968)

LIST OF KENTISH TOWN BUTCHERS SHOPS; 1766-2010

The following list with the exception of branch shops outside the area comprises the majority of retail butchers shops that have opened for business in Kentish Town. The dates given are not definitive, but merely indicate a shop of that name/title was recorded at the address stated. Where a shop came into existence under the old system of numerous sub-division names for stretches of property in the same road or street the modern equivalent road /street name has been given. In instances where a boundary road extends beyond or only one side forms part of Kentish Town discrepancies will arise.

ALLCROFT ROAD
84 A. Swingler 1874

BRECKNOCK ROAD (part)
H. Lee 1858
4 John Twigg 1899
18 William H. Attfield 1881-1884
18 Henry Pearce 1893-1901
18 Aubrey Guyer 1910-1940
18 P. F. Cramer c1960
18 Corrigan Bros 1969-2010
35 Harry Barnes 1893-1910
35 Henry R. Rushworth 1915
35 & 37a, G. W. Cager's 1923
35 Cyril Isles 1940
229 George J. Davis 1893-1899

CAMDEN ROAD (part)
63 A. & .A. Whitton Ltd 1940-1960
63 D. Doyle 1969
63 Corrigan Bros 1970-1980

CARLTON STREET
77 Charles Cato 1862-1867
77 William Lynes 1881
77 Robert Corbett 1884-1885
77 Thomas Burkett & Co 1893-1895
77 Richard A. Bayler 1899-1901

CASTLE ROAD
35 John W. Harris 1852-1862
70 William Morgan 1867

CASTLEHAVEN ROAD
4 Henry Burrows 1881
13 Thomas Barrett 1855-1857
28 Frederick Bishop 1862
32 William Searle 1884-1885
52 Richard Pyle 1855-1882

CHALK FARM ROAD
Thomas Weedon 1841
10 Walter Whitlam c1850-1915
10 Leonard P. Hillman 1920
10 Stanley Kite 1934-1972
26 William Harbone 1893
31 Charles Bulpin 1877
31 H. Walsingham 1878-1910
31 Frances Wise 1912-1915
63 George Frost 1882
67 Thomas McCague 1910

CHETWYND ROAD
Charles Cato 1881
69 Richard Selway 1901-1938
69 George Thomas 1963
69 Corrigan Bros 1969-1975
69 Jackson Bros 1989
69 M. Ghosseiri [T/A Jackson] 1989-2010

CROGSLAND ROAD
64 M. Jakings 1884-1885
64a James Thomas Anwyl 1893-1899
64a William Folks 1904
64 Edward H. Ellicote 1910

DARTMOUTH PARK HILL (Left Side)
7 Joseph Bryant 1885
7 Alfred H. Cansick 1940
127 George Ryan 1885
141 William G. Harman 1940
141William G. Ivin 1940-1969
147 Mrs M. Kempton 1940-1969
147 Corrigan Bros 1975
149 Thomas Willis 1885
16 C. Elgar & Sons Ltd 1969 (Road)

FERDINAND STREET
Hexmore 1858
13 James Keeling 1855
14 James Hale c1860-1884. TD
26 William Eltlton 1862-1867
26 William Dale 1881-1885
32 Edmund & Alfred Davies 1852-1895
36 Charles W. Hobbs 1893-1895
40 W & J Davey 1855 -1867
40 Walter Jones 1872
47 B.A. Davis 1893-1899
62 Godfrey R. Gluck 1899
62 John Golterboth 1903
62 Frederick Englhard 1904
62 John Ulm 1912-1915
62 Edward A. Wallington 1931
62 Harry M. Sinclair 1940

FORTESS ROAD
1 George Cooper 1885
1 Berry Butchers 1855
2 John French 1867-1874
7 T.P. Christofides 1969
24 Thomas Pitcher 1901-1911
24 Thomas Hurrey 1912-1920
24 Charles H. Pike 1940
24 Pike & Woolridge 1952
53 Henry Hurst 1899-1903
61 Lakis Meat Products 2004
111 S. Simons & Co 1881-1885
111/ 151 William Wooldridge 1893-1899
116 T.R. Lawrence & Sons 1937
116a Thomas Musgrove 1895-1910
116a William G. Roney 1912
116a Henry T. Lee 1913
116a Ernest F. Vincett 1913-1969
116a B.H. Fox 1992
116a Benny Butchers 1994-1995
122 Charles Harry Rouch 1895-1911
122 Ludwig Goetz 1915
132 Robert Wilson 1893-1903
132 Wheeler Bros 1910-1915
151 Henry Albert Gayes 1903-1904
151 Wheeler Bros Sons 1910 -1940
151 J.H. Dewhurst Ltd 1960
151 T.W. Downs Ltd 1969

GILDEN ROAD
30 George F. Wells 1910
33 Thomas Meaby 1874

GRAFTON ROAD
Keats [1 Railway Arch] 1866-1867
John Death [1 Railway Arch] 1868
23 Richard Nickles 1862-1867
William Turner [Railway Arch] 1872
50 George J. Davis 1884-1885
78 Alfred R. Bayler 1899-1901

HARMOOD STREET
30 Robert Gate 1862-1867
30 Robert Elvidge 1872-1896
47 Milton Hersant 1855-1896

HAWLEY ROAD

29 T.W. Jones 1862-1882
29 Thomas Barnett 1863
29 Chas H. Carr 1893

HIGHGATE ROAD

Phillip & Edward Wilson 1805
Joseph Holland 1805
2 David Berry 1862
4 John & Wm Chas Spink 1881-1882
4 George Cooper 1884-1885
5 Mrs Hannah Figures 1841-1852
77 Edward Davis 1861-1874
75/77 William O'Hara 1881-1896
75/77 Lidstone Ltd 1896-1912
81 Lidstone Ltd 1950
81 L.E. Tucker 1969
105 Meat Boutique 1994
123 Lidstone Ltd 1899-1910
137 Edward Davis 1872
157 John French 1867-1915
157 J.H. Ritchie 1940-1969

ISLIP STREET

100 William Plaistow 1882 - 1883

KELLY STREET

2 William Satterly 1881
2 J. Balfour 1940

KENTISH TOWN ROAD

William Jepson
Joseph Rippington
Phillip Arrowsmith c1800
William Hill 1813
R & S Silverside 1823-1837
Frances Kemp 1839
Christopher J. Palles 1840
Edward W. Dober 1841
Thomas King 1841
Richard Bussan 1846
Benjamin Buckle 1846
11 William Edycoomb 1851
11 James Holt 1855-1857
10a H. Heller 1950-1960
10a R.A. Hyde 1969
14 William Pain 1851
16 William Chipperfield 1851-1852
16 Edward Hook 1867
17a John Hawes 1895
17a David Clowson 1910
17a Jack 1913-1915
17a Joseph Heller 1940
67 H & E Hook 1862-1882
67 Philip Stone & Sons 1883-1903
67 William Wallace 1901-1912
67 David H. Churchouse 1913-1954
67 R & T Danvers 1958
67 William Daniels 1960
100 William Cooper 1864-1877. TD
117 Charles Davies 1867-1893
117 Robert Elvidge 1881-1912
120/122 London Co-operative Society 1932-1954
122 Central Catering 1964-1970
131/133 J.H.Ritchie 1910 -1969
141 Batchelor Bros 1899
144 Guy Walter Smith 1893
144 James Henry Balch 1895-1903
144 Rolfe & Tasch 1926
155 Thomas Josling 1855-1867
155 Stephen Langbridge 1877
155 Frances Richardson 1883-1885
155 William Gould 1893-1915
155 John Hunter & Sons 1940
155 J.H. Ritchie 1954-1969
159 Caleb Edwards 1872
160 G. W. Thompson c1860-1881. TD
160 Thomas C. Hale 1895-1913. TD
172 Henry Robinson 1877
172 Tom Durrant 1881-1883
172 Joseph Thwaite 1890-1895
172 M.L.W. Redhead 1899-1901
172 Charles Dray 1902-1904
172 Cecil Grovener Williams 1906

172 Percival N. Brazil 1913
186 John Hunter & Sons 1913-1950
186 J. H. Ritchie 1954-1958
188 Appleyards 1878-1912
190 Josiah Fardoe 1862
190 J. Thomas 1867
190 Henry Eli Read 1868-1883
190 William J. Morton 1895
190 H. Ritchie 1898-1969
196 George Sanders 1895
196 John Edwards 1898-1915
206 Mrs C. Knight 1867-1915
208 Matthew C. Knight 1870-1915
216 English & Colonial Meat Co, 1903
216 S. Simons & Co 1903-1910
217 P. Degruttola 1994-1995
222 James Preston 1910-1915
222 West (Butchers) Ltd 1920-1960
222 West Layton Ltd 1963-1969
222 J.H. Dewhurst Ltd 1975-1995
223 Robert A. Balch 1862-1914
225 Thomas Hale 1766
249 Frank Sanford 1893
249 Tom Crisp & Co 1895
249 Thomas Burkett & Co 1899-1903
258 B & M Seafoods [Pure Meat Co] 1995-2010
258 Pure Meat Co 2002-2006
260 George Price 1910-1915
280 Ernest Nicklinson 1890-1901
280 Valentine Lunch & Co 1902-1904
287 Charles Davies 1823-1877
287 Thomas Boreham 1882
287 Henry Love 1890
287 Frederick Turner 1892-1895
287 Harry Fox 1896
313 J.H. Dewhurst Ltd 1930-1965
317 Richard S. Morris 1862-1867
317 Walter Hughes 1872
317 Henry J. Hook 1877-1904
317/319 G.F. Kimber 1907-1955
317/319 J.H. Ritchie 1955-1965
387 William Golding 1872
387 James Grant 1877

LEIGHTON ROAD

Thomas Williams 1855
4 Alfred Bennett 1874
4 Adolphos Attwell 1881-1882
4 Pond & Hawkins 1884
18 Anthony Salt 1855
89 William Cornish 1862-1882
89 William O'Hara 1890-1896
89 Lidstone Butchers Ltd 1896-1901

LISMORE CIRCUS

5 Thomas Islip 1867
7 Charles Batho 1862-1867
16 William Packer 1874-1885
16 Frank Crease 1901-1915
16 William T. Hagan 1931
16 William Mailing 1941
16 H. Gilry 1969

MALDEN ROAD

6 Robert Carter 1861-1877
14 Herbert Ellicott 1913
25 Chas Pitt 1867-1874
25 Mrs J. Good 1877
25 Thomas Greenslade 1882-1883
37 James Miller 1855-1883
37 Robert Elvidge 1890-1912
51 William Keen 1922-1926
51 Keen & Vincett 1927-1929
51 Ernest R. Vincett 1929- 1964
51 P & G Butchers 1969
54 Charles Summerlin 1862-1867
54 George Vincett 1871-1883
54 William Webb 1895
54 James Butler 1901
54 Thomas Price 1927-1928
61 Henry Webb 1855-1926
61 Horace Tipple 1927
61 Thomas W. Buttling 1928-1971
87 Charles Cato 1877-1883
87 Robert Gate 1882 (!)

97 John Holford 1862-1877
106 Thomas Price 1931
110a J. K. Buer & Sons 1940
147 John Williams 1863-1864
147 Luke P. Major 1867-1870
147 Charles Cato 1874-1877
147 Charles Hurst 1877-1883
147 Frederick J. Neal 1884-1886
154 Henry Loscome 1881-1883
159 Benjamin Judge 1867-1877
159 Henry Curle 1881-1922
159 Henry Granger 1926-1931
159 D. Smith & Sons 1938-1966
159 Patrick's Butchers 1969
159 W. White 1974

MANSFIELD ROAD
70 Wilson Bros 1931
70 Leslie Hutchings 1940-1958
76 Samuel Brown 1889-1903
76 Arthur Heathfield 1910-1915
117 Berry Butchers 1881-1889
119 Nash & Collins 1899-1903
119 Thomas Knight 1915

PRINCE OF WALES ROAD
11 Alfred Grey 1867-1870

PRINCE OF WALES CRESCENT
3 Mr Holland
3 W & M Hersant 1862-1899
3 Samuel Groose 1901
3 Henry Goldfinch 1910
3 Charles H. Euinton 1931
3 Bartlett Bros 1940
23 W & Edward Hook 1855-1872

QUEENS CRESCENT
14 Henry J. Honiball 1854-1883
56a Alfred Keen 1881-1883
56 Rayner, King & Hardwick 1910
58a George Anderson 1881-1885
58 Towers, Rayner & King 1893-1899
58 W.F. Hardwick 1899
58 North Western Meat 1903
58 Rayner Ltd 1922-1935
58 H. Denby Ltd 1950-1958
58 Central Stores Butchers 1960
58 R & C Butchers 1960-1974
58 Economy Meats 1978
70 Towers, Rayner, King 1877-1884
72 Anne Farey 1866
72 John Markham 1869
72 J. Jackson 1874-1876
72 Thomas Hale 1877-1884. TD
74 Rayner & King 1874-1912
74 James Rayner Ltd 1912-1934
74 Thomas W. Buttling 1935-1986
74 D. Cole 1992-2003
76 Thomas C. Hale 1891-1927
76 Hale & Piper 1938-1954
76 May & Shepperd 1954-1959
76 George Bayliss 1960-1984
82 George Lacey 1867
82 John Evans 1869
82 Edward Andrews 1874-1914
82 Preston & Co Ltd 1915
82 West (Butchers) Ltd 1922-1935
86 J. Allen 1867-1870
86 Joseph D. Thorn 1874-1893
86 Towers, Rayner & King 1895-1901
86 Towers, Rayner, Hardwick 1905
86 Rayner Ltd 1906-1922
88a Ernest Nicklinson 1895-1899
88a Metropolitan Meat Co 1903-1904
88a Fritter & Co 1908
90 P.N. Brazil 1895-1915
90 Frederick P. Bonham 1922-1954
90 Bobs Butchers 1955-1992
102 Harry W. Hobbs 1893
102 Owen Walters 1895
102 Stone & Roberts 1899-1905
102 Thomas Stone 1905-1910
102 Beard & Bailey 1922-1929

131 John Holford 1872
131a Horace J. Woolton 1950-1958
131a Drake (K. T.) Ltd 1960-1974
131a Mr Meat 1984-1995
131a Martin Croucher1995-1997
139/141 London Co-operative Society 1960-1969
143 Frederick Jones 1867. TD
145 T. Nickles 1867-1870
149 Henry Davis 1882-1883
149 Francis Grabham 1932-1933
149 J.H. Ritchie 1938-1997
155 Robert R. Warren 1915-1923
155 F. C. Shelly 1923-1827
155 Isaac Franks 1929-1940
167 Van Mingeroet 1954 HFD
167 D. King 1958-1960
167 Martin Stone 1960-1962

REGIS ROAD
24/27 Fairfax Meadow plc. 1989 - 2010

RHYL STREET
26 Thomas Mundy 1915
26 Thomas Bailey 1928

ROCHESTER PLACE
5a Thomas Jack 1940

ROYAL COLLEGE STREET (part)
11 Henry Follet 1862-1867
55 Frances King 1855
79 H. Collingwood 1862-1868
79 Mrs Anne Attfield 1895
111 Mrs D. & W. Nickles 1855
114 Mrs D. Nickles 1867-1874
114 Chas Silk 1877
114 John Stone 1872-1901
116 Mrs D. & W. Nickles 1867
120 George Bryce 1885
122 Silvanus Webber 1915
148 Corrigan Bros 1975-1992
158 Charles W. Nadauld 1915
166 William J. Harris 1870-1896
166 John Penny 1899-1904
166 Silvanus Webber 1915
166 C. Gold 1969
176 T. Fardon 1867
223 Richard Bros 1867
223 J.H. Bowbeer 1874
223 Edmund C. Trueman 1893-1901
223 Joseph S. Bryant 1910

SOUTHAMPTON ROAD
39 Walter Clark 1889-1899
39 Joseph & Albert Watkins 1910-1940
39 Walter White 1960-1969

SWAINS LANE
9 Cavours 1940-1997
21 Silvanus Webber Ltd 1940-1969
21 Elite Meats 1985-2010

TORRIANO AVENUE
4 Alex Thorn 1855
59 A.G. Rogers 1940-1969
64 Martins Catering Butchers
73 John Oldfield 1862
94 John Tappin 1862-1872
94 James Hawker 1881-1892
94 Frederick Harman 1893
94 Henry N. Berkley 1895-1910

WARDEN ROAD
4 Edmund C. Wood 1862
7 Robert Roberts 1881-1899
7 John Hales 1910

WEEDINGTON ROAD

49 James Honie 1862-1864
49 W.M. Turner 1867-1872
49 & 51 James Wooley 1877
49 George H. Tye & Son 1884-1885
49 Thomas Holland 1888
51 J. Richardson 1862-1864
51 John Seeney 1895
65 W. B. Fallover 1868
164 Robert T. King 1867

WELLESLEY ROAD

52 Mrs C. Kirkland 1881-1915
66 George Cox 1872

Abbreviation
TD = Tripe Dresser
HFD = Horse Flesh Dealer

Index

A

B

C

D

E

F

G

H

I

J

K

L

S

T

U

V

W

Y

Lightning Source UK Ltd.
Milton Keynes UK
173539UK00001B/22/P